Year 4 Practice Book

What did you do in maths in Year 3?

Draw or write what you enjoyed doing most.

This book belongs to Raayah.

My class is Mars.

Contents

It is time to start!

How to use this book

Do you remember how to use this **Practice Book**?

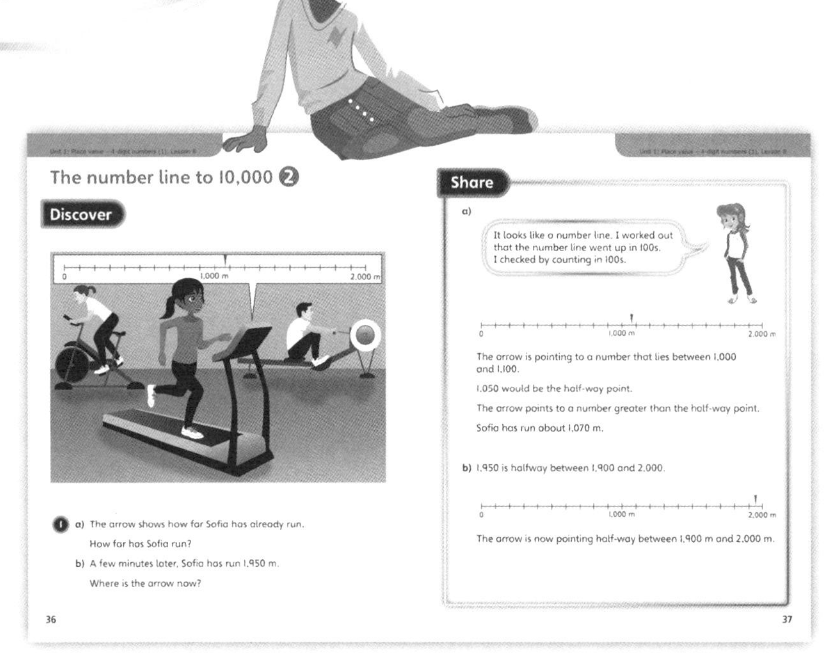
Unit 1: Place value – 4-digit numbers (1), Lesson 8

The number line to 10,000 2

Discover

1 a) The arrow shows how far Sofia has already run. How far has Sofia run?

b) A few minutes later, Sofia has run 1,950 m. Where is the arrow now?

Share

a)

It looks like a number line. I worked out that the number line went up in 100s. I checked by counting in 100s.

0 1,000 m 2,000 m

The arrow is pointing to a number that lies between 1,000 and 1,100.

1,050 would be the half-way point.

The arrow points to a number greater than the half-way point.

Sofia has run about 1,070 m.

b) 1,950 is halfway between 1,900 and 2,000.

0 1,000 m 2,000 m

The arrow is now pointing half-way between 1,900 m and 2,000 m.

36 37

Use the **Textbook** first to learn how to solve this type of problem.

Unit 1: Place value – 4-digit numbers (1), Lesson 7 → Textbook 4A p32

The number line to 10,000 1

1 What number is being shown on each number line?

a) 0 1,000 2,000 10,000

The number [ ] is shown.

b) 4,000 4,100 4,200 5,000

The number [ ] is shown.

c) 3,200 3,210 3,300

The number [ ] is shown.

24

This shows you which **Textbook** page you need.

Have a go at questions by yourself using this **Practice Book**. Use what you have learnt.

Challenge questions make you think hard!

Questions with this light bulb make you think differently.

Reflect

Each lesson ends with a Reflect question so you can think about what you have learnt.

Use My Power Points at the back of this book to keep track of what you have learnt.

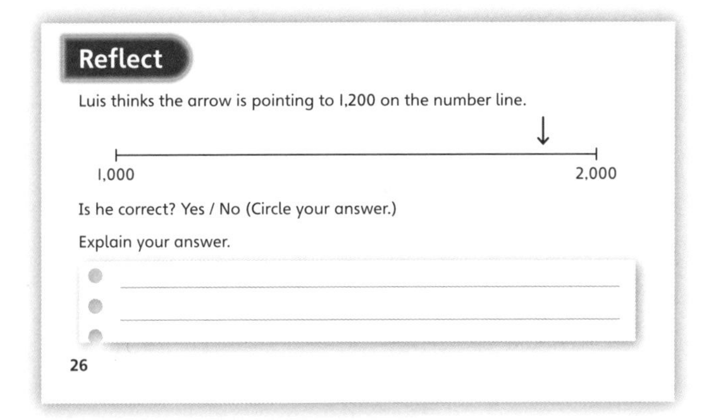

Reflect

Luis thinks the arrow is pointing to 1,200 on the number line.

1,000 — 2,000

Is he correct? Yes / No (Circle your answer.)

Explain your answer.

26

My journal

At the end of a unit your teacher will ask you to fill in My journal.

This will help you show how much you can do now that you have finished the unit.

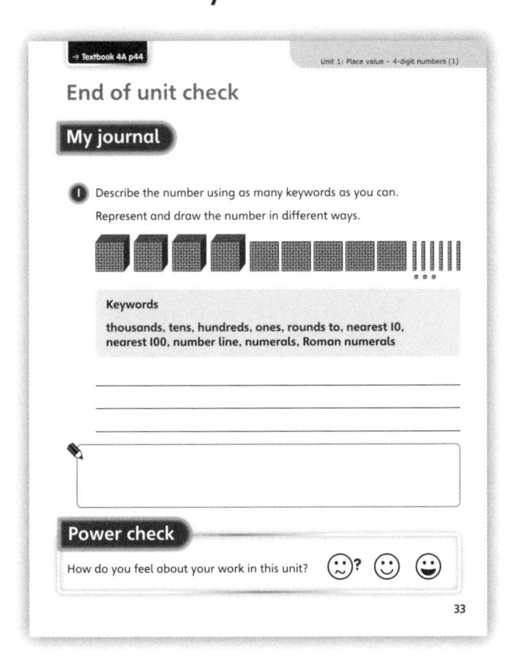

→ Textbook 4A p44

Unit 1: Place value – 4-digit numbers (1)

End of unit check

My journal

1. Describe the number using as many keywords as you can.
Represent and draw the number in different ways.

Keywords
thousands, tens, hundreds, ones, rounds to, nearest 10, nearest 100, number line, numerals, Roman numerals

Power check

How do you feel about your work in this unit?

33

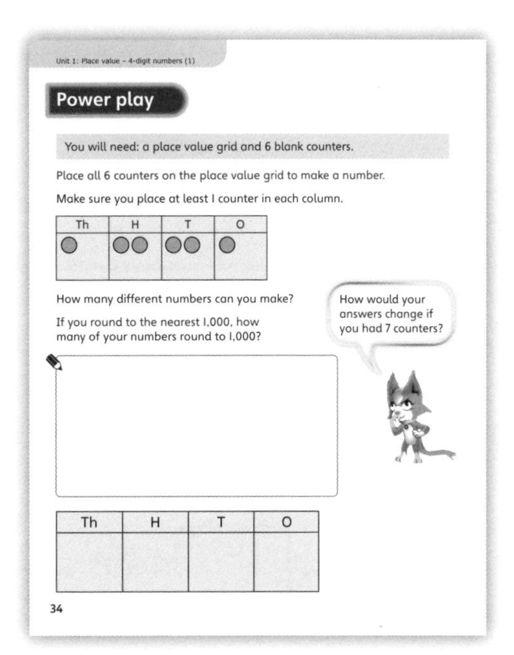

Unit 1: Place value – 4-digit numbers (1)

Power play

You will need: a place value grid and 6 blank counters.

Place all 6 counters on the place value grid to make a number.

Make sure you place at least 1 counter in each column.

Th	H	T	O
●	●●	●●	●

How many different numbers can you make?

If you round to the nearest 1,000, how many of your numbers round to 1,000?

Th	H	T	O

34

→ Textbook 4A p8

Numbers to 1,000

1 Write the numbers represented in words and numerals. One has been done for you.

Representation	Words	Numerals
	Three hundred and sixty-three	363
100 100 100 100 100 10 10 10 10 10 10 10 1		
100 100 100 100 100 100 10 10 10 10 10 10		
1 0 8		

2 Partition the numbers.

a) 892 = ☐ hundreds, ☐ tens and ☐ ones

b) 705 = ☐ hundreds, ☐ tens and ☐ ones

3 Jamilla is drawing numbers.

Use the key to complete the numbers.

Key

□ = 100 ▯ = 10 ○ = 1

a) 721 =

H	T	O

b) 204 =

H	T	O

c) 330 =

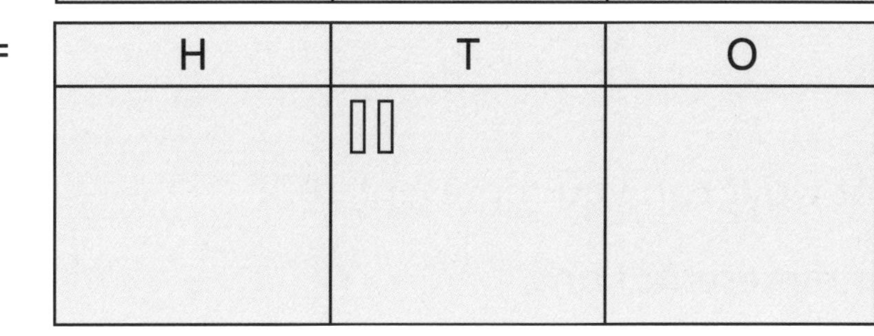

H	T	O

4 Richard is using base 10 equipment to make a number.

Richard says, 'I have made 513.'

Is Richard correct? Explain your answer.

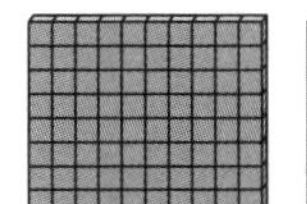
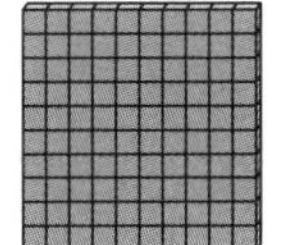
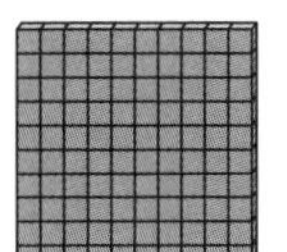
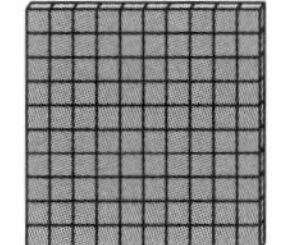
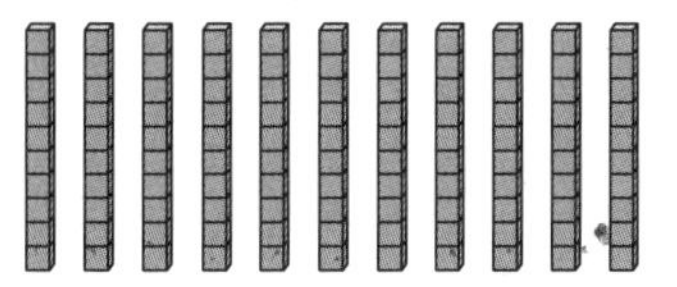

Richard is not correct because there are 400 11 tens and 3 unit's so the answer is 413

5 Mo is making four hundred and fifty-two.

CHALLENGE

100 100 100 10 10 10 10 1

Mo has dropped some place value counters. Which counters might he have dropped?

Is there more than one answer?

____

____

____

Reflect

How many different ways can you represent 707?

Make or draw your representations.

Explain your representations to a friend.

→ Textbook 4A p12

Rounding to the nearest 10

1 **a)** Sort the numbers into the table.

Round down to the nearest 10	Round up to the nearest 10
333 41 902 102 981	209 15 78 457

b) Write two extra numbers in each box.

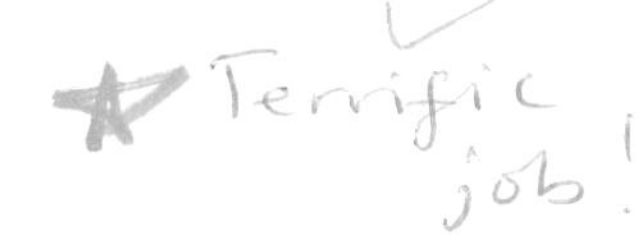

2 **a)** Complete the number lines and round the number shown to the nearest 10.

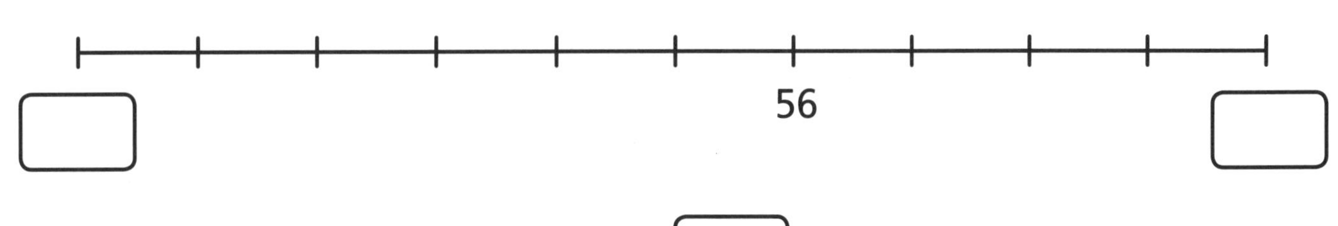

56 rounded to the nearest 10 is ☐.

b)

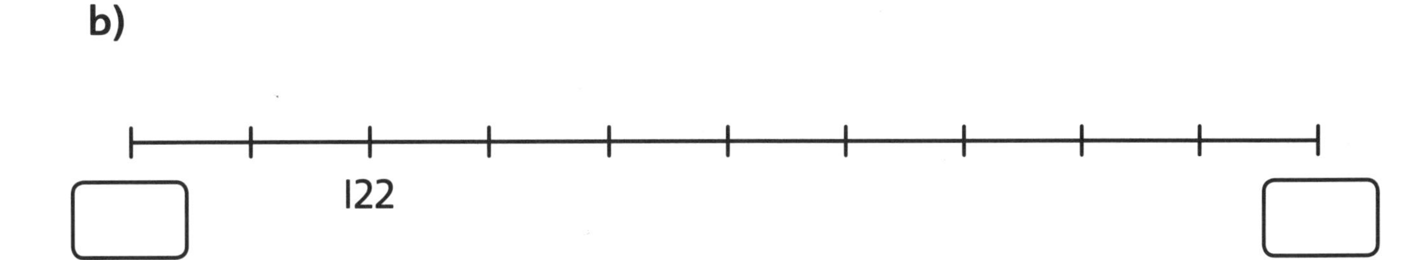

122 rounded to the nearest 10 is ☐.

3 Kim's age rounded to the nearest 10 is 20.

What is the oldest Kim could be?

Kim could be ☐.

I wonder if Kim could be older than 20.

4 Complete the number sentences.

a) 18 to the nearest 10 is ☐.

b) 28 to the nearest 10 is ☐.

c) 48 to the nearest 10 is ☐.

d) 78 to the nearest 10 is ☐.

e) 98 to the nearest 10 is ☐.

f) 128 to the nearest 10 is ☐.

g) 368 to the nearest 10 is ☐.

5 Round the numbers to the nearest 10.

a) 76	80	176	180	376	380
b) 234	230	367	~~36~~ 370	45	40
c) 450	450	711	716	98	100

CHALLENGE

6 Two numbers both round to 50, to the nearest 10.

The total of the two numbers is 99.

What could the two numbers be?

Mark them on the number line below.

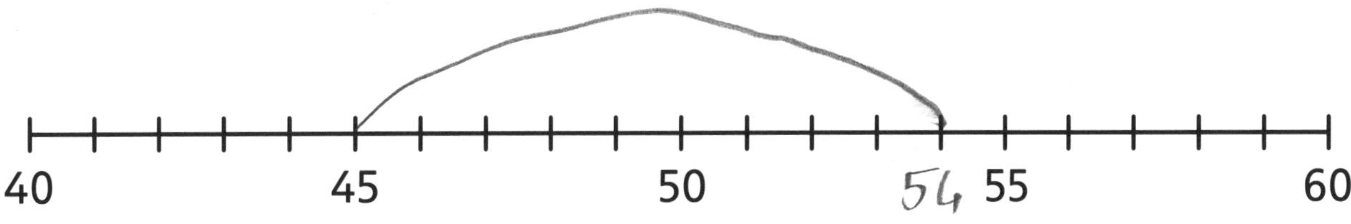

I think I can find more than one answer. I will start by picking a number that rounds to 50.

Reflect

Hannah has a 2-digit number.

7

The first digit is 7.

Can Hannah decide if she should round the number up or down to the nearest 10? Why? Why not?

→ Textbook 4A p16

Rounding to the nearest 100

1 Round the numbers to the nearest 100.

a)

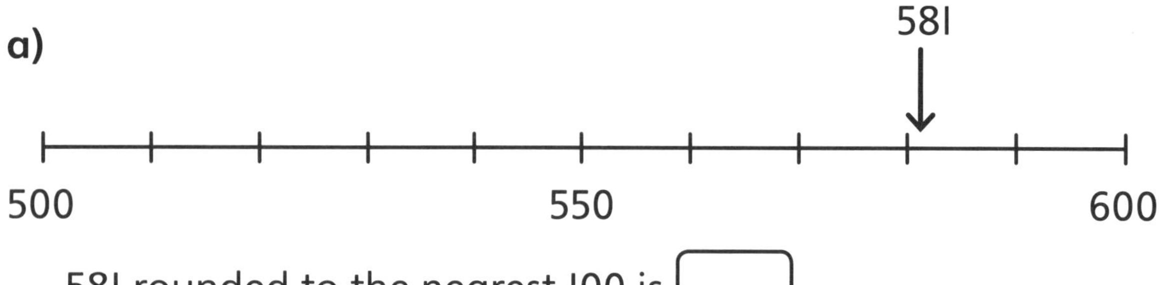

581 rounded to the nearest 100 is ☐.

b)

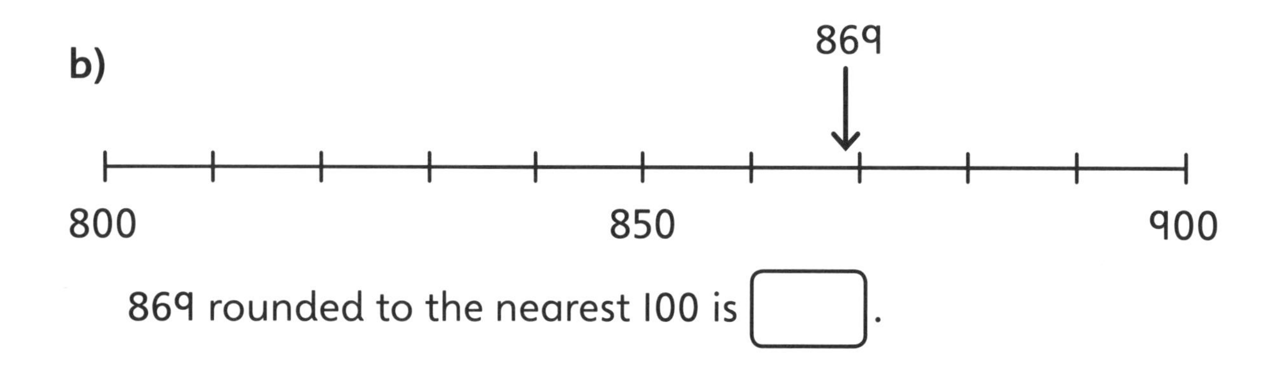

869 rounded to the nearest 100 is ☐.

c) Mark three hundred and nine on the number line.

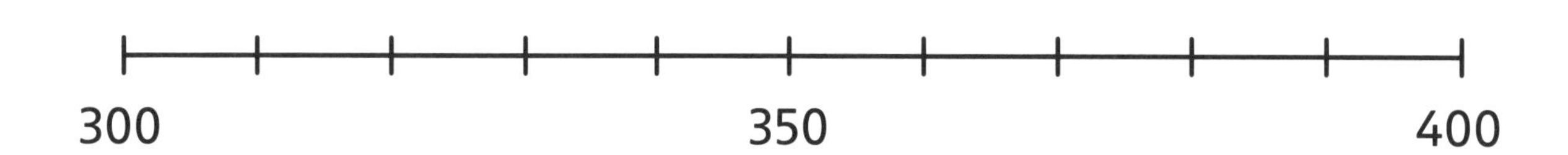

Three hundred and nine to the nearest 100 is __________.

2 Show and write three numbers on the number line that round to 400 to the nearest 100.

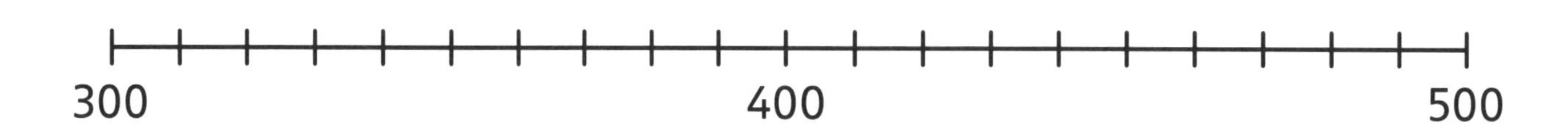

3 Circle the numbers that will round **up** to the nearest 100.

740 698 652 378 412 301 746

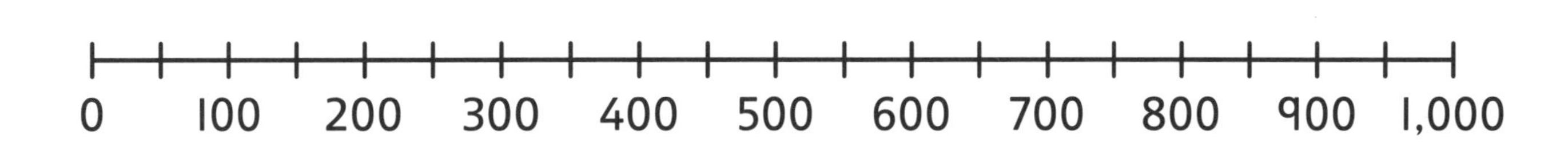

4 Draw lines to match the numbers to the rounded numbers.

Number | **Rounded to nearest 100**

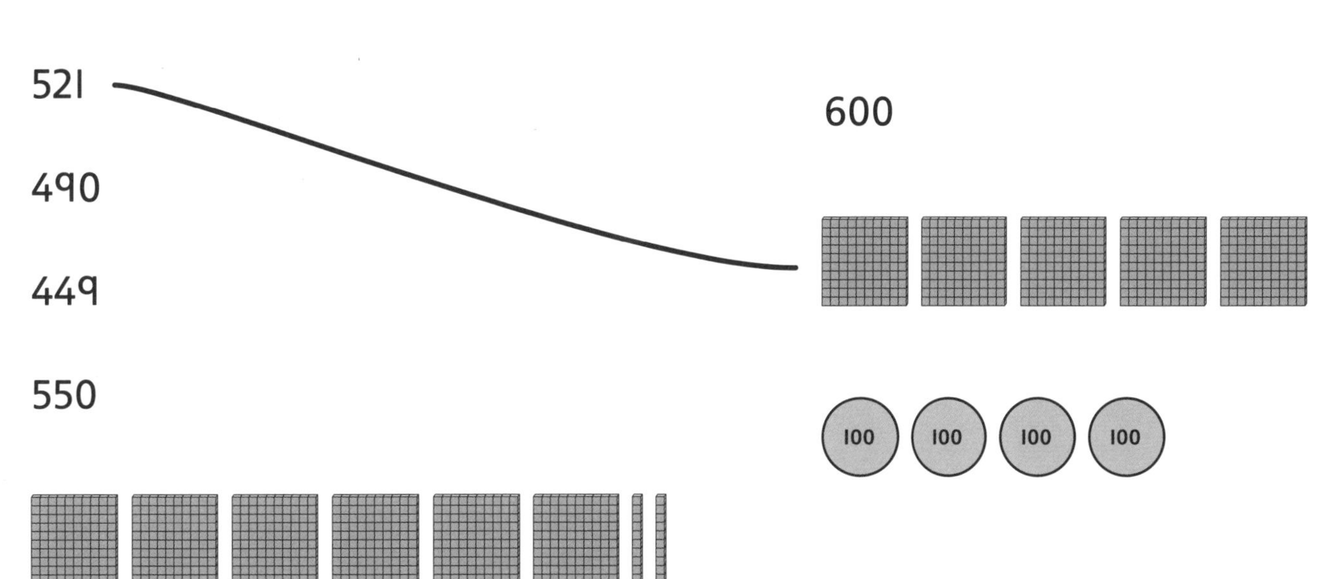

5 Round the numbers to the nearest 100.

a) 768 ☐

b) 402 ☐

c) 199 ☐

d) 84 ☐

e) 951 ☐

f) 12 ☐

g) 420 ☐

250

Bella says 250 can be rounded to either 200 or 300 because it is in the middle of these numbers.

Richard disagrees and says it rounds down to 200.

Do you agree with either Bella or Richard? Explain your answer.

Reflect

Describe how you would round 462 to the nearest 100.

→ Textbook 4A p20

Counting in 1,000s

1 How many cups are there altogether?

Write your answers in numerals and words.

a) 

There are [ ] cups.

There are ______________________________ cups.

b) 

There are [ ] cups.

There are ______________________________ cups.

c) 

There are [ ] cups.

There are ______________________________ cups.

2 Complete the number tracks.

a)

2,000	3,000			6,000		8,000

b)

0		2,000		4,000		

c)

10,000	9,000			6,000		4,000

3 a) What numbers are being represented? Write your answers in numerals and words.

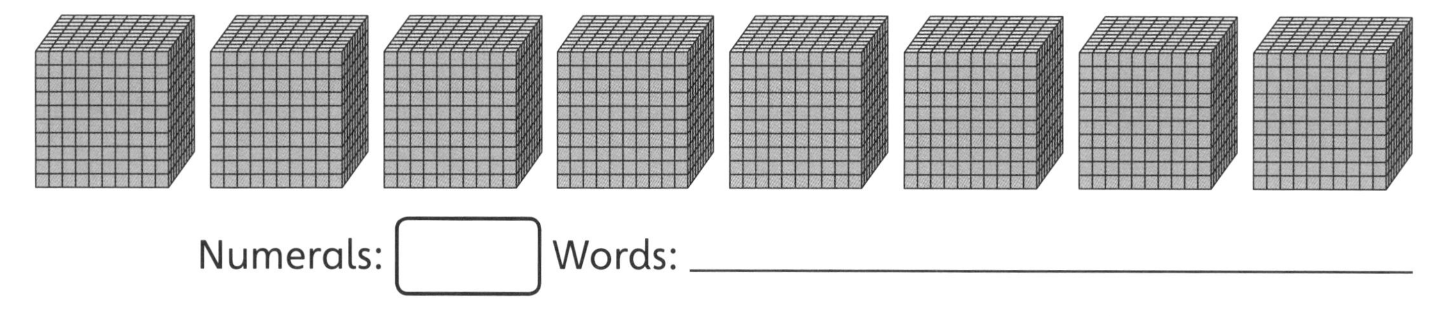

Numerals: ☐ Words: ______________________

b)

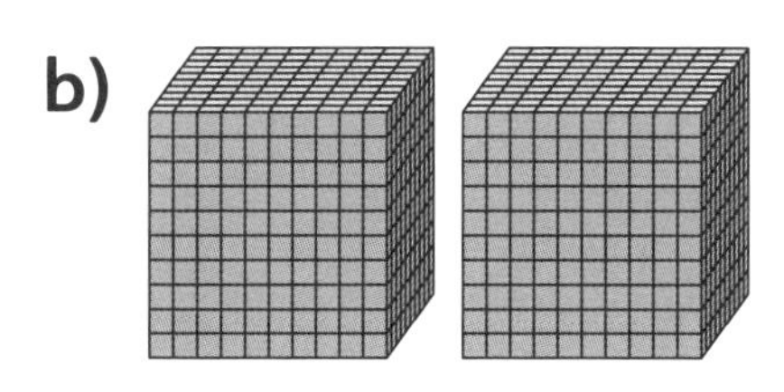

Numerals: ☐ Words: ______________________

4 Draw three thousand in base 10 equipment.

5 What number is being represented? Explain your answer.

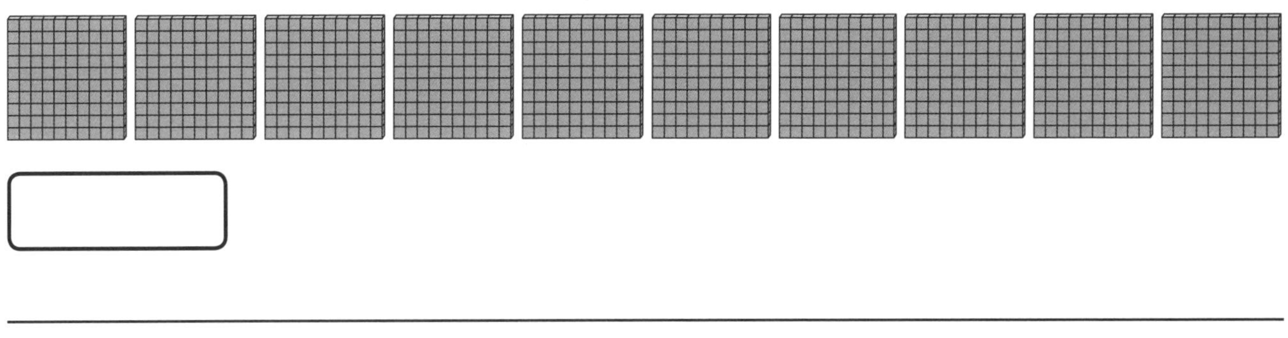

6 Andy counts from 2,000 to 7,000 in 1,000s.
At the same time Bella counts down from 6,000 to 3,000 in 1,000s.

Will they ever say the same number at the same time?
Use a number line to explain your answer.

Reflect

2,000 pencils are red and 5,000 pencils are blue. The rest are green.

Show or explain how you can work out how many green pencils there are.

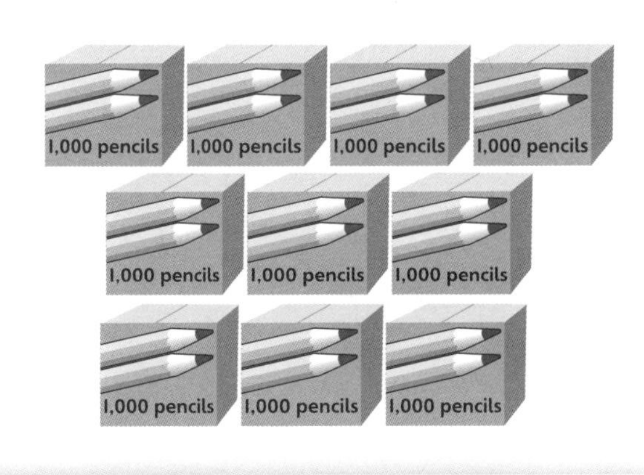

→ Textbook 4A p24

Representing 4-digit numbers

1 Write the numbers represented in numerals and words.

Base 10 equipment	Numerals	Words
		________ thousand, ________ hundred and ____________

2 Complete the part-whole models.

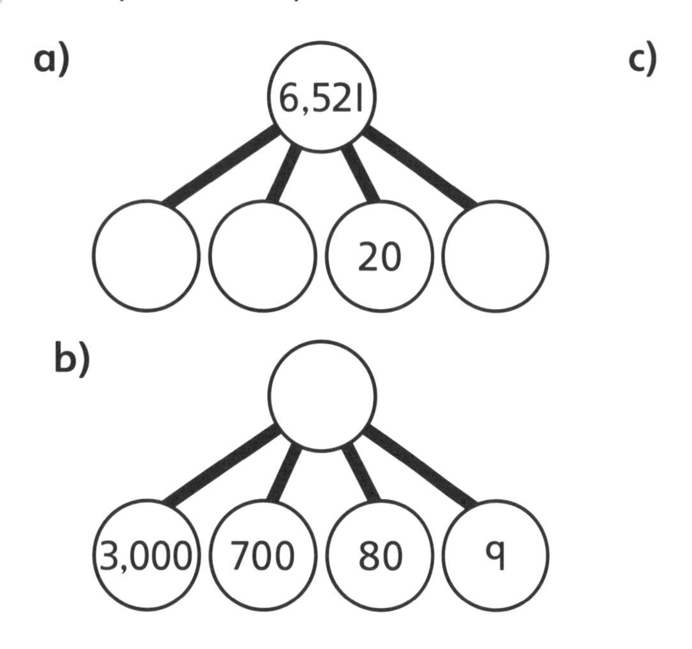

c)

3 Draw lines to match the base 10 equipment to the correct part-whole model.

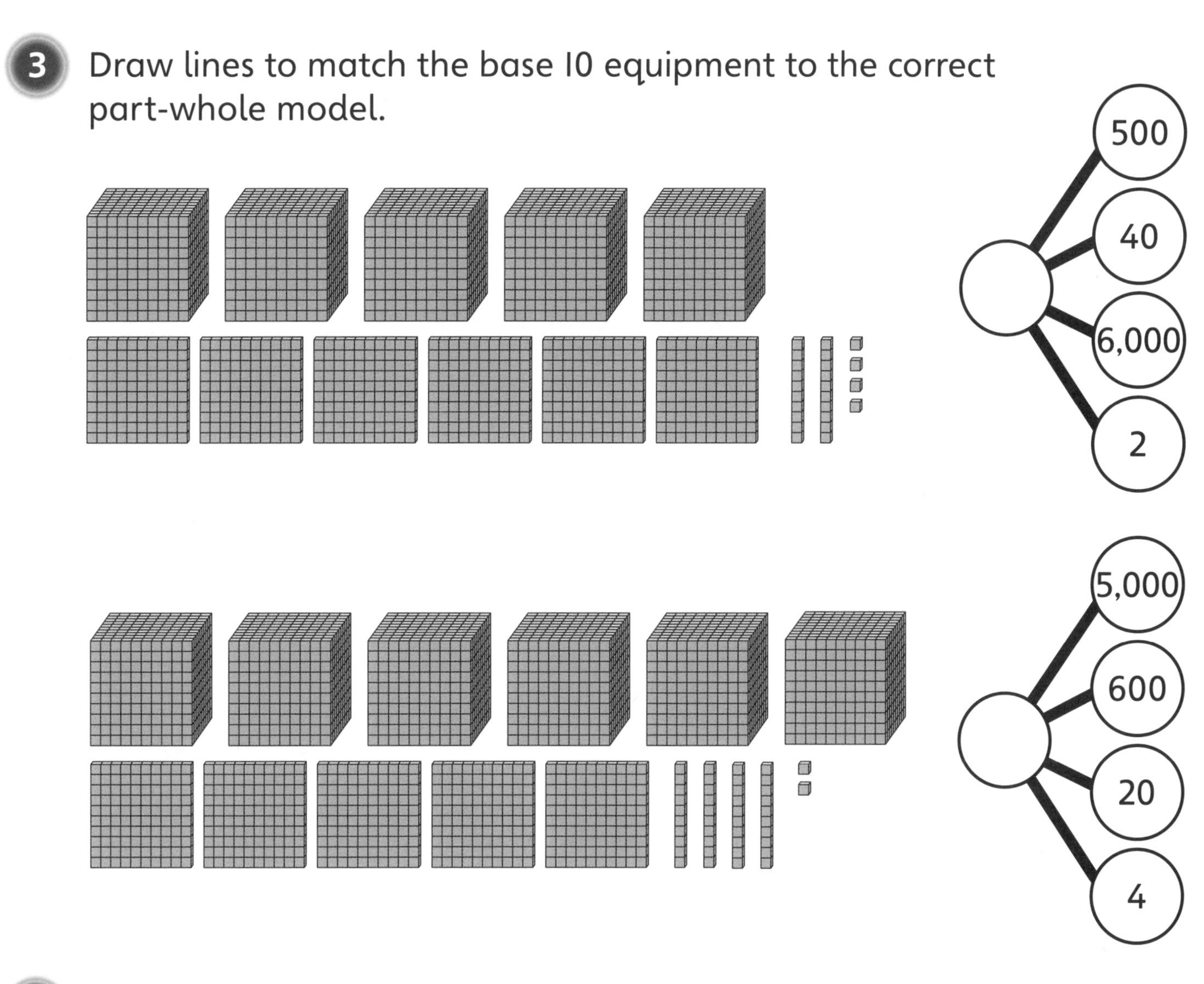

4 Which of the three numbers represented is different from the other two?

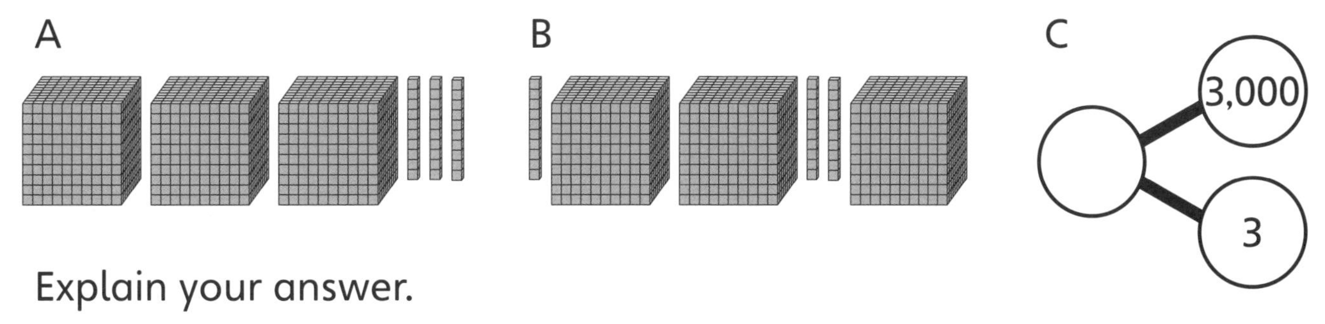

Explain your answer.

5 Reena has made a number.

CHALLENGE

She thinks she has made a 3-digit number.

Do you agree? Yes / No (Circle your answer.)

Explain your answer.

____

____

____

Reflect

Choose four digit cards to make a number.

Represent the number using base 10 equipment and a part-whole model.

→ Textbook 4A p28

1,000s, 100s, 10s, and 1s

1 Write in the number that is represented.

a)

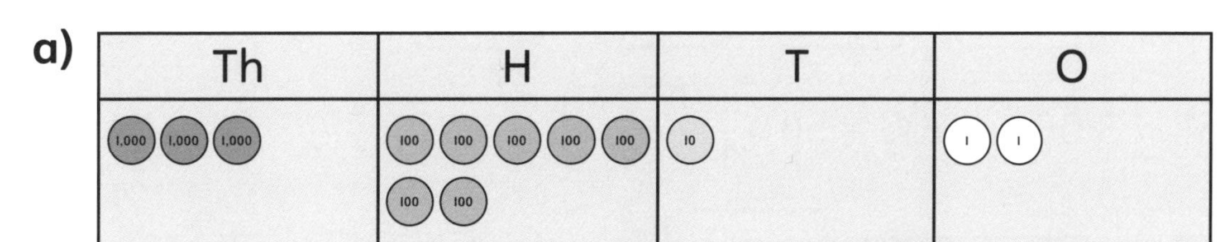

b)

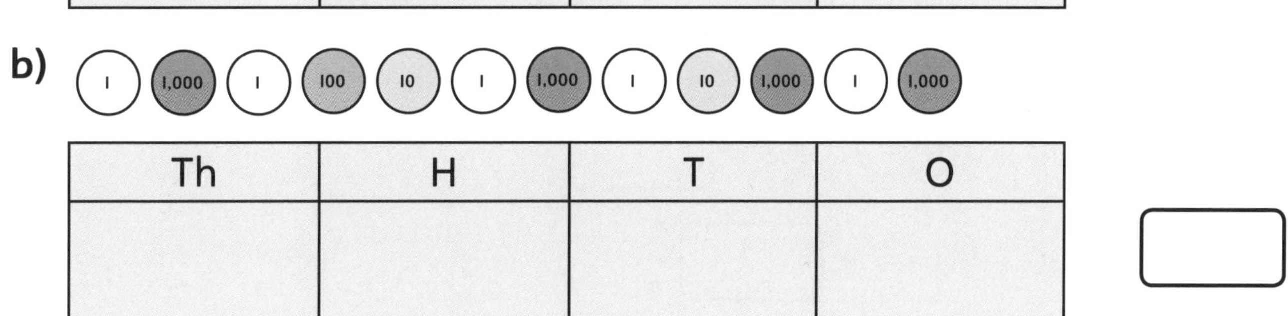

Th	H	T	O

2 Draw counters on the place value grid to represent each number.

a) 2,356

Th	H	T	O

b) Four thousand, eight hundred and four

Th	H	T	O

c) 2,000 + 200 + 50 + 6

Th	H	T	O

3 Complete to make the number sentences correct.

a) 3,458 = ☐ + ☐ + ☐ + ☐

3,458 = 3,000 + 400 + 50 + ☐

3,458 = 400 + 3,000 + ☐

b) 3,000 + 700 + 70 + 2 = ☐

3,000 + 50 + 7 = ☐

3,000 + 500 + 70 = ☐

4 Which one is the odd one out? Explain why. 

a) 3,000 + 1,700 + 40 + 9

b) Four thousand, seven hundred and forty-nine

c)

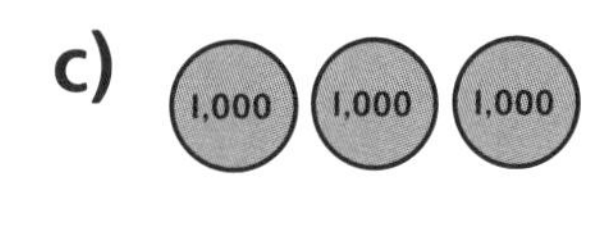

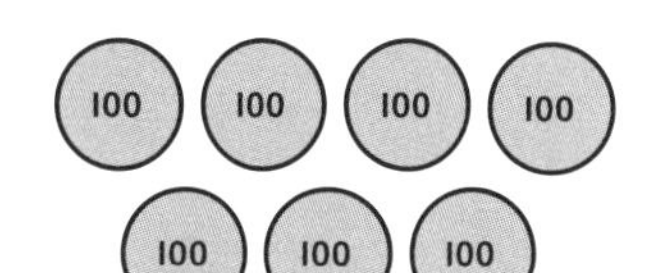

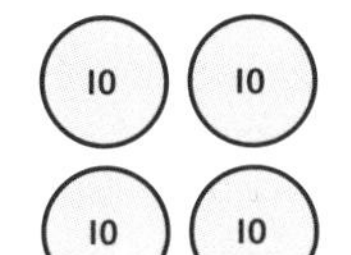

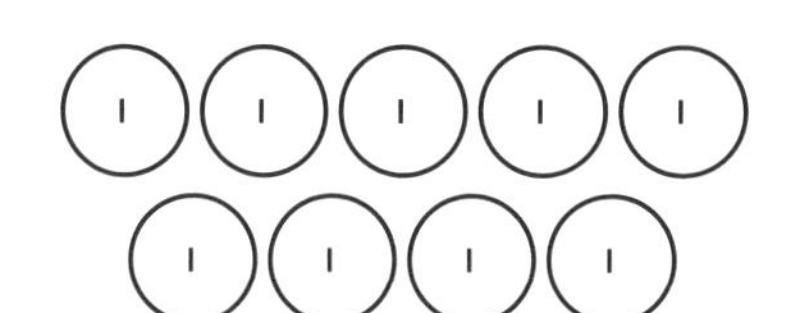

d)

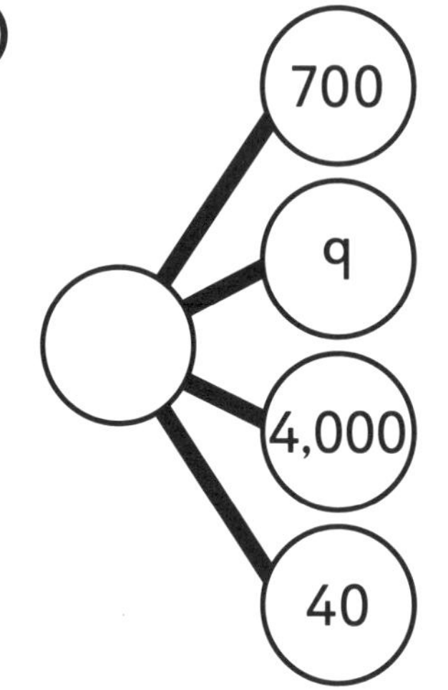

____

____

____

____

____

CHALLENGE

5 Andy has made a number. He says:

- 'My number has the same number of 1,000s and 10s.'
- 'There are two more 1s than 10s.'
- 'The hundreds digit is half the thousands digit.'

What could Andy's number be?

Draw place value counters to show the possible answers.

Reflect

Mr Harris says, 'In this village there are 2,300 people.'

Mrs Mackintosh says, 'There are 23 streets here with 100 people in each street.'

Explain why they can both be correct.

→ Textbook 4A p32

The number line to 10,000 

1 What number is being shown on each number line?

a)

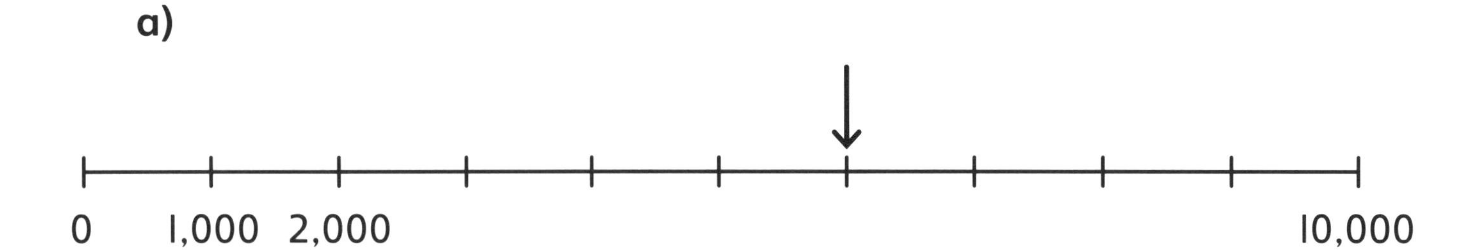

The number ☐ is shown.

b)

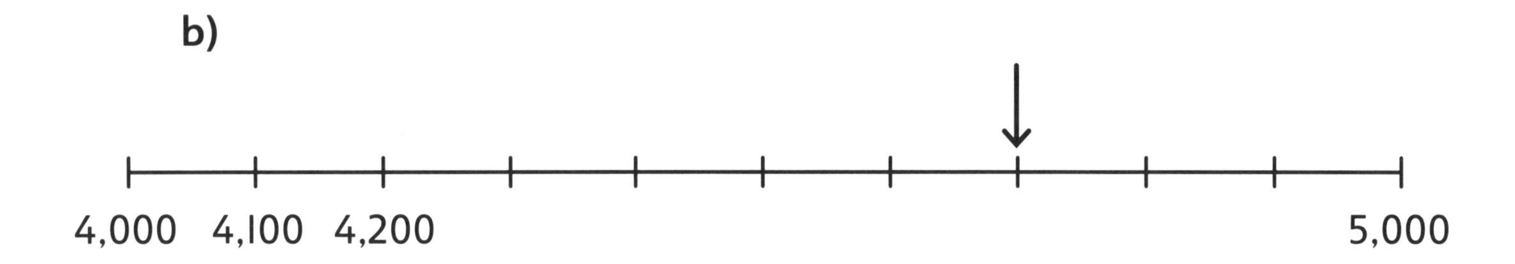

The number ☐ is shown.

c)

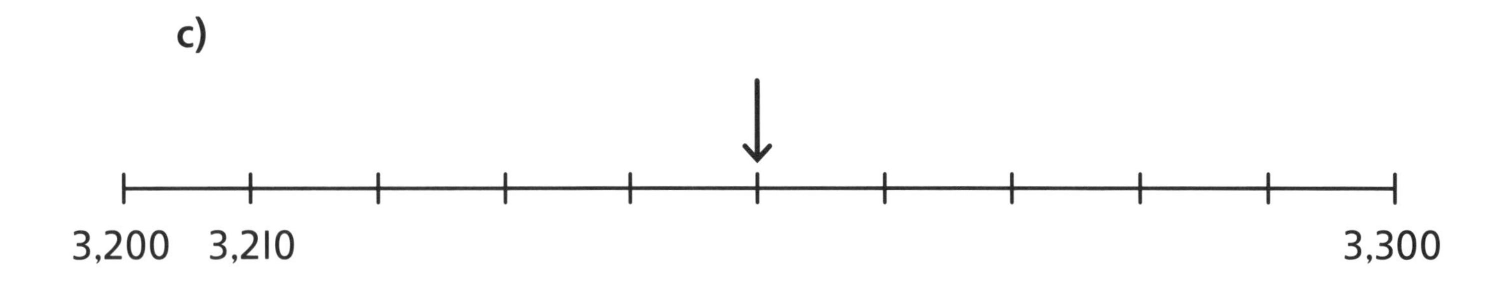

The number ☐ is shown.

2 Write the missing numbers.

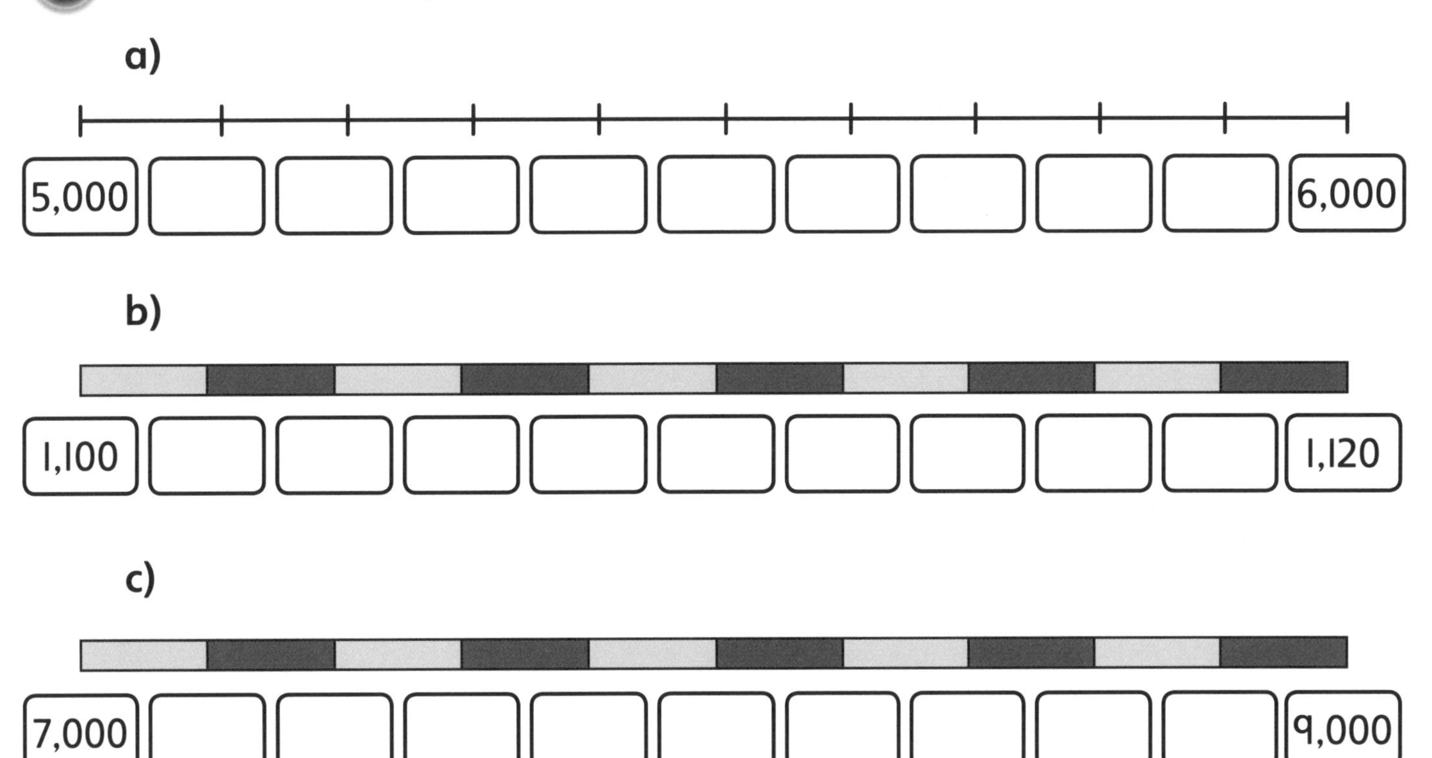

3 Draw an arrow to show each number on the number line.

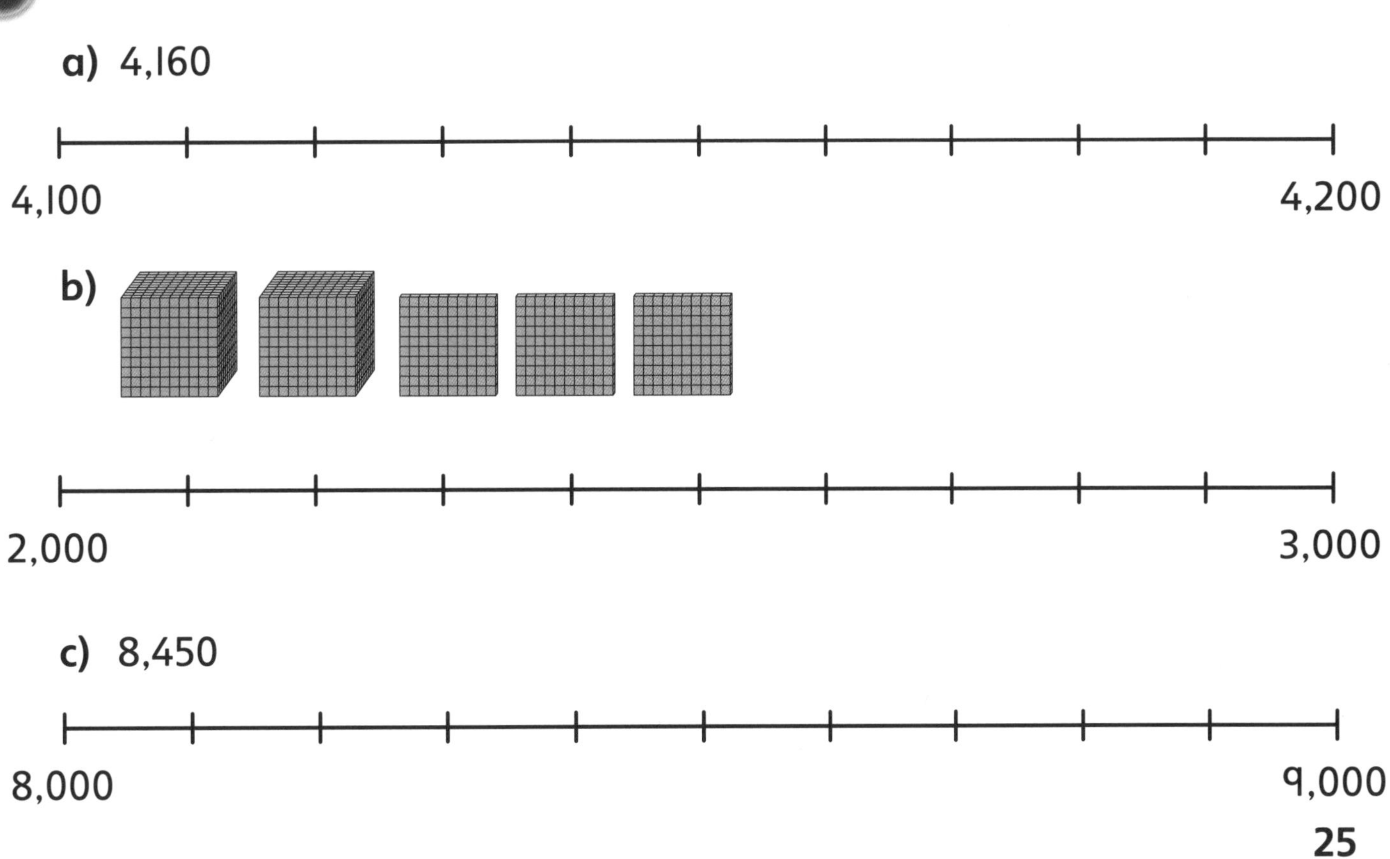

4 **a)** Max says that arrow A is at 1,200. Explain why Max is incorrect.

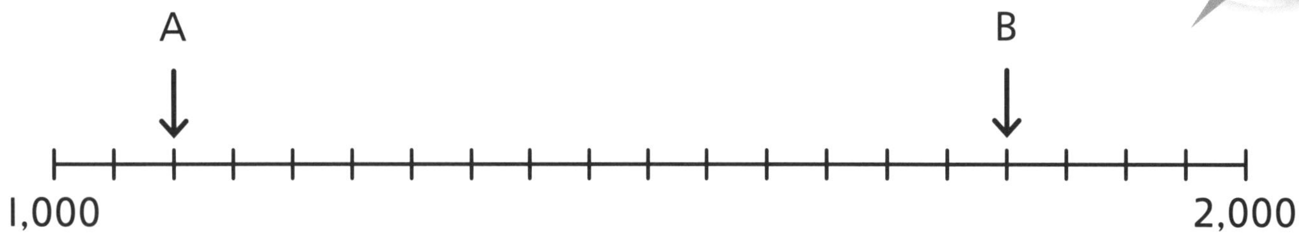

____

____

____

b) What number is arrow B pointing to?

Arrow B is pointing to ☐.

Reflect

Luis thinks the arrow is pointing to 1,200 on the number line.

1,000 2,000

Is he correct? Yes / No (Circle your answer.)

Explain your answer.

____

____

→ Textbook 4A p36

The number line to 10,000 2

1 Show the numbers on the number lines provided.

a) 7,800

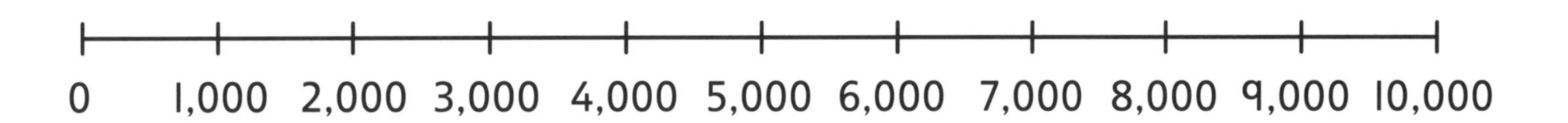

b)

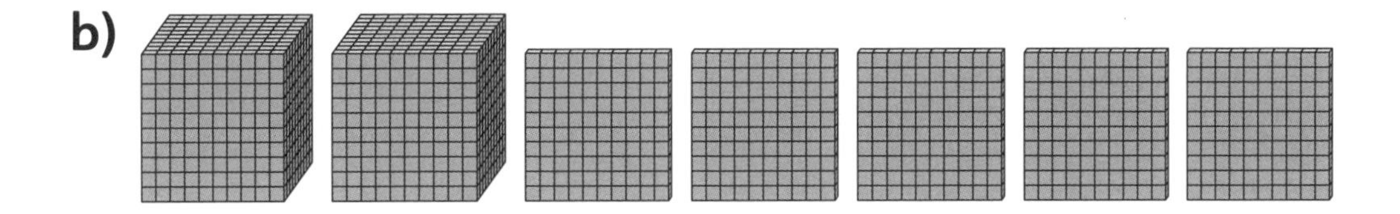

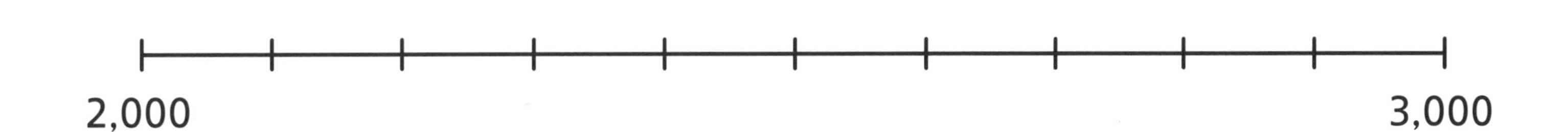

c) Four thousand four hundred

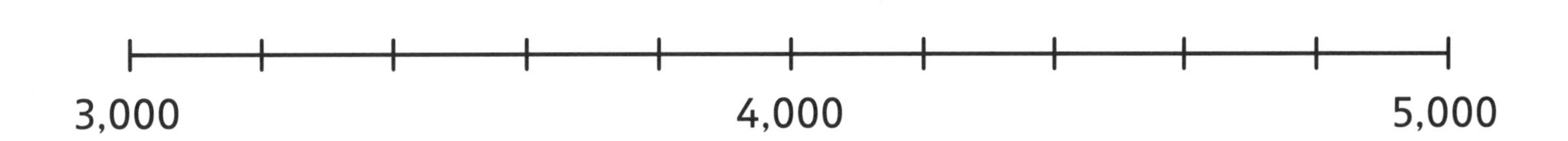

2 Show 4,600 on each of the number lines.

a)

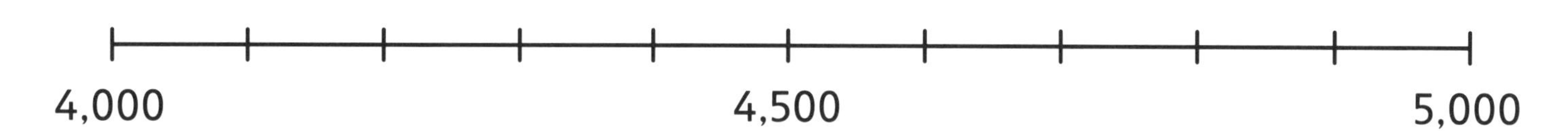

b)

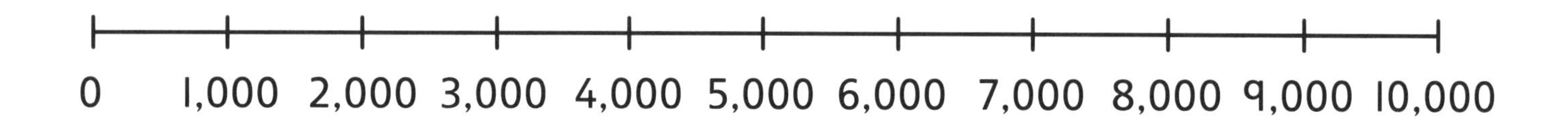

3 Write three numbers that could appear on each number line.

a)

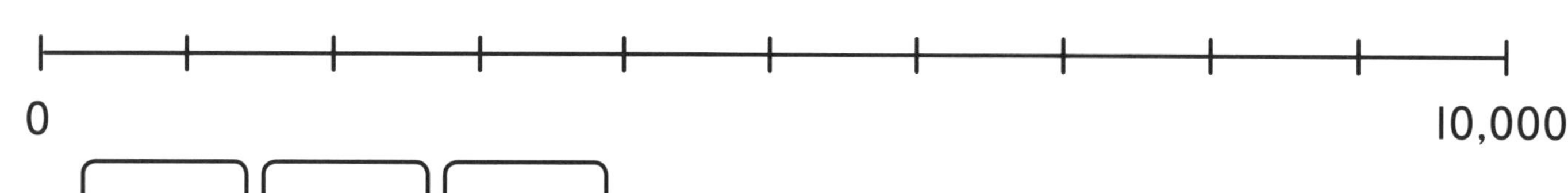

b)

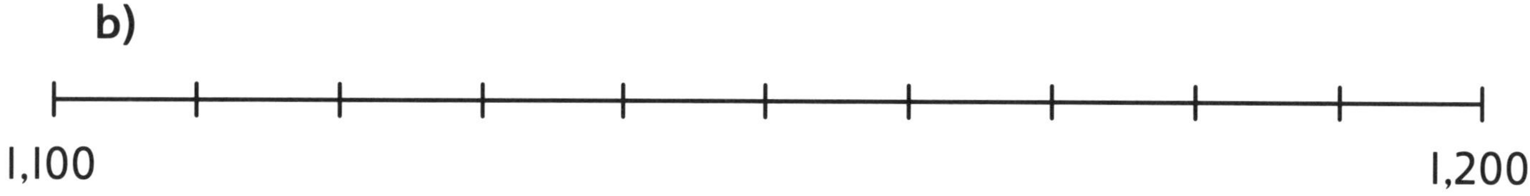

c)

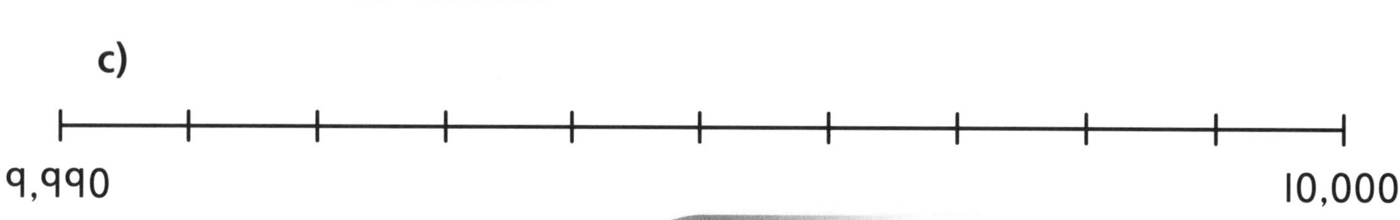

I will work out the intervals on the number line first.

CHALLENGE

4 Estimate what numbers the arrows are pointing to at A, B and C.

A B C

1,000 3,000

A = ☐ B = ☐ C = ☐

Reflect

The following numbers can be placed on the number line.

4,250 6,790 3,425 5,400

Which numbers could go at each end of the number line?

Explain your answer.

I think there is more than one answer.

→ Textbook 4A p40

Roman numerals to 100

1	I	11	XI
2	II	12	XII
3	III		XIII
4	IV		XIV
5	V	15	XV
6	VI	16	
7	VII	17	XVII
8	VIII	18	XVIII
9	IX	19	XIX
10	X	20	

50	L
100	C

1 Which numbers do these Roman numerals represent?

a) 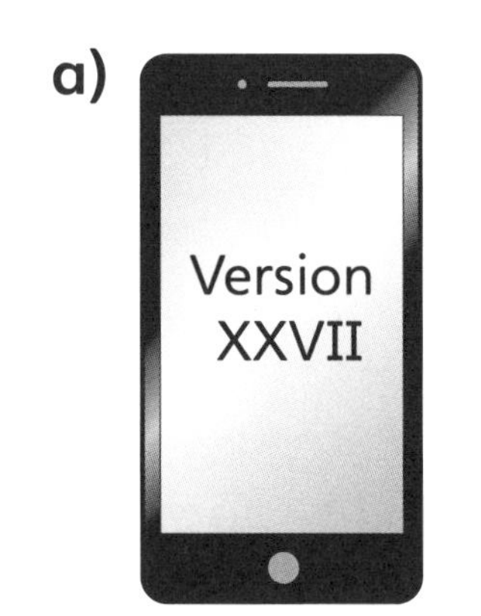

b) 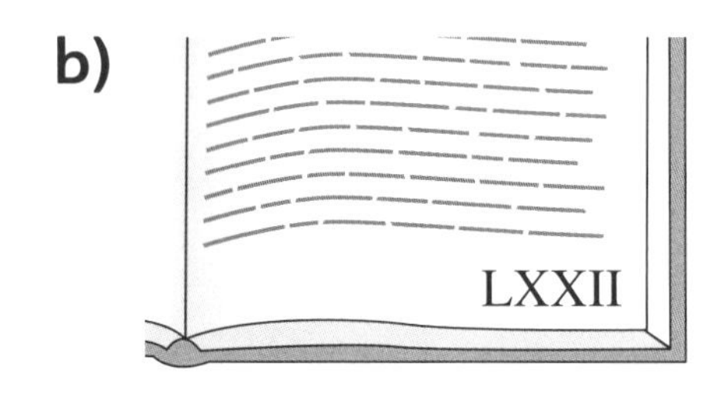

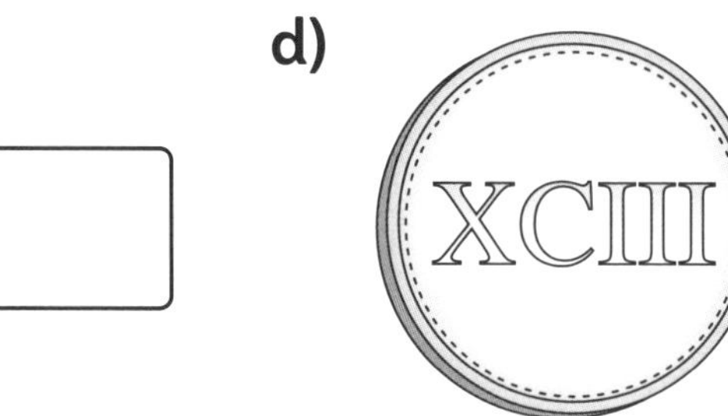

c) 

d) 

2 What times are shown on the clocks?

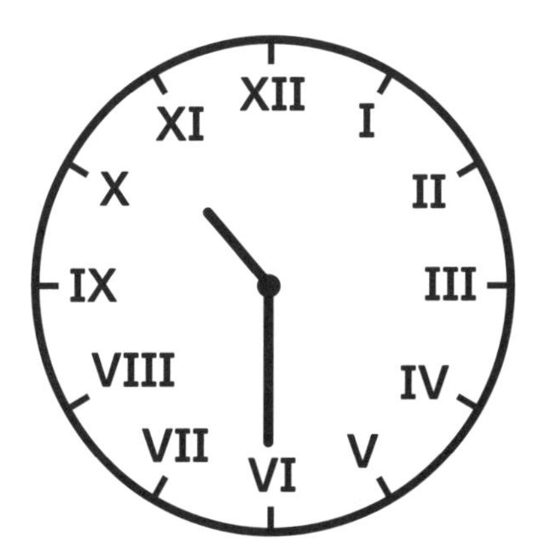

____________________ ____________________

3 Draw lines to match up the cards. Then fill in the missing number and Roman numeral.

XXII		LXXIV	XXXVI 	LXIX	LXXXIV	XCIX
99	36	42	22	84		69

4 **a)** What is V more than XXI? ☐ 

b) Emma has XXXIII marbles.

Amelia has LXX marbles.

Who has more marbles? How many more does she have?

_______________ has more marbles.

She has ☐ more than _______________ .

5 Complete the number sentences.

a) LX + ☐ = C

b) XXXII + ☐ = C

c) X × X = ☐

d) C – ☐ = XIV

e) L + L = ☐ + XXXV

Reflect

Circle the numbers greater than 50.

XXV L LI XL XLV LXXI

I think there is a way of doing this without working out all of the numbers.

How did you work out which numbers to circle?

→ Textbook 4A p44

End of unit check

My journal

1 Describe the number using as many keywords as you can.

Represent and draw the number in different ways.

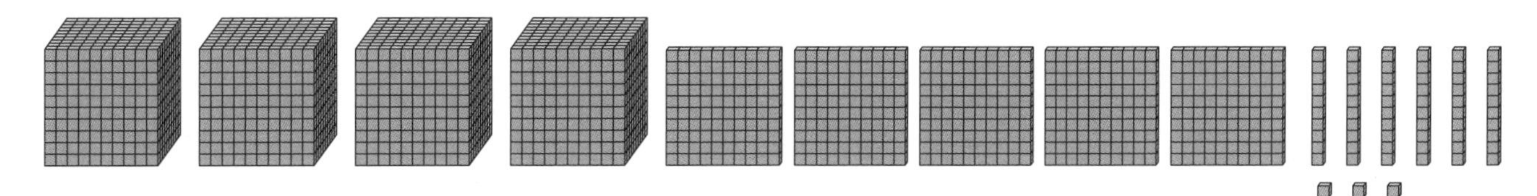

Keywords

thousands, tens, hundreds, ones, rounds to, nearest 10, nearest 100, number line, numerals, Roman numerals

Power check

How do you feel about your work in this unit?

Power play

You will need: a place value grid and 6 blank counters.

Place all 6 counters on the place value grid to make a number.

Make sure you place at least 1 counter in each column.

Th	H	T	O
●	●●	●●	●

How many different numbers can you make?

If you round to the nearest 1,000, how many of your numbers round to 1,000?

How would your answers change if you had 7 counters?

Th	H	T	O

→ Textbook 4A p48

Finding 1,000 more or less

1 **a)** What is 1,000 more than:

Th	H	T	O

1,000 more than ☐ is ☐.

b) What is 10 more than:

Th	H	T	O
1,000 1,000 1,000 1,000 1,000	100 100 100 100 100 100 100 100	10 10 10 10 10 10 10	

10 more than ☐ is ☐.

c) What is 100 more than:

Th	H	T	O
1,000 1,000	100 100 100 100 100 100 100 100 100	10 10 10 10 10	

100 more than ☐ is ☐.

d) What is 100 less than:

Th	H	T	O
1,000 1,000 1,000 1,000 1,000 1,000 1,000 1,000 1,000 1,000			

100 less than ☐ is ☐.

2 Write the number shown by each representation, then complete the table.

Number	Number in digits	1,000 more	100 less	10 more
1,000 1,000 1,000 1,000 100 100 100 100 1 1 1 1 1 1 1				
0 500 1,000 1,500 2,000 2,500 3,000				
Seven hundred and fifty-eight				

3 Fill in the missing numbers to make the sentences correct.

a) 1,000 more than 4,879 is ☐ .

b) 100 less than 4,879 is ☐ .

c) 10 more than ☐ is 4,879.

d) 1 more than 4,879 is ☐ .

e) 3,921 is 1,000 more than ☐ .

f) 100 less than ☐ is 652.

4 Complete each number sentence.

a) [2,885] more than 2,875 is 2,876.

b) 5,783 + [1000] = 6,783

c) [ ] less than 3,580 is 3,480.

d) [ ] – 10 = 3,990

e) 5,999 + 1,000 – 10 = [ ]

f) [ ] + 10 – 100 = 7,860

g) 7,500 is [ ] less than 8,500.

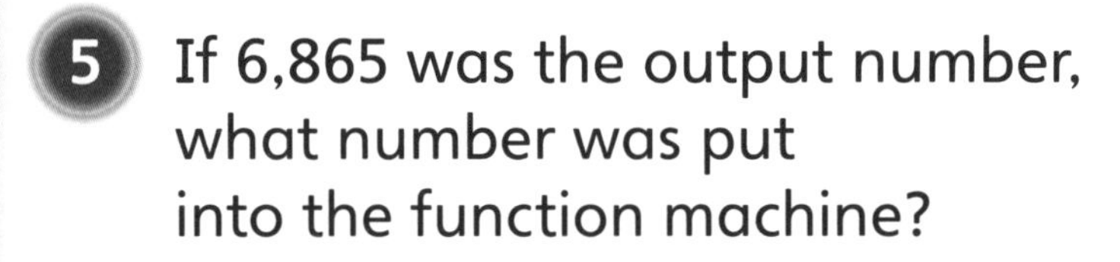

5 If 6,865 was the output number, what number was put into the function machine?

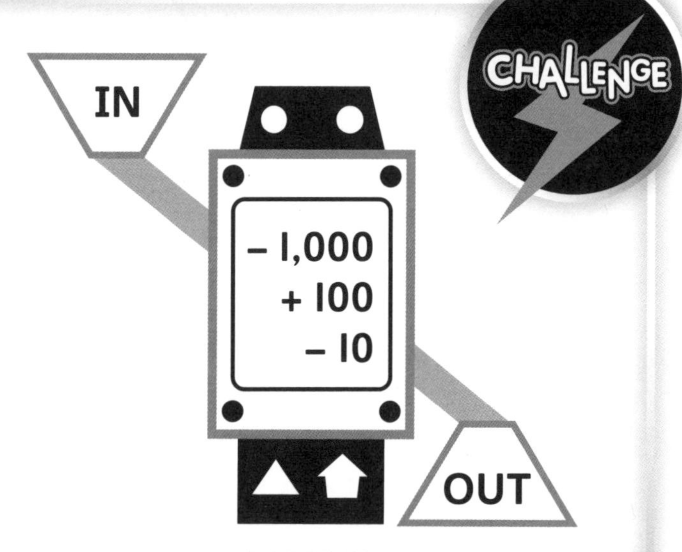

The number [ ] was put into the function machine.

Reflect

When finding 1,000 more or 1,000 less than a 4-digit number, which digits change? How many digits change?

→ Textbook 4A p52

Comparing 4-digit numbers 1

1 Complete using **more than**, **less than** or **equal to**.

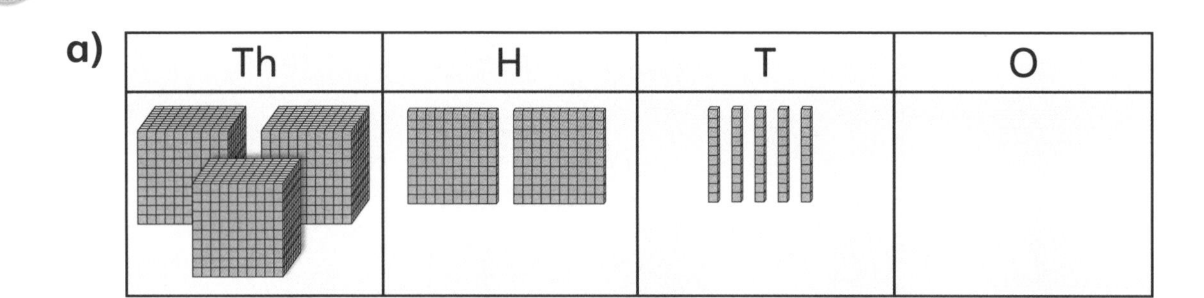

a)

Th	H	T	O

is ______________________________

Th	H	T	O

b)

Th	H	T	O

is ______________________________

Th	H	T	O

2 Complete using <, > or =

a)

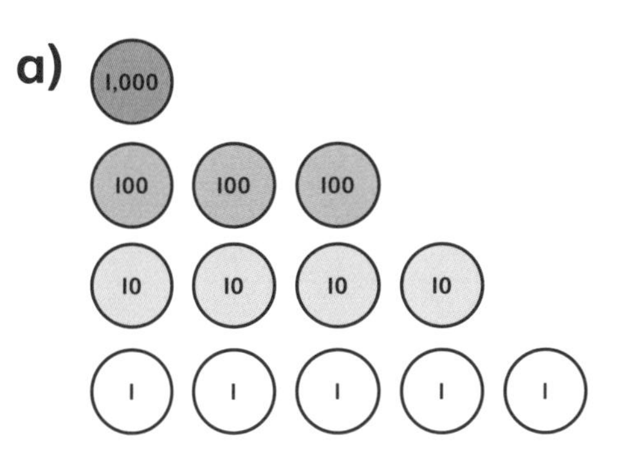

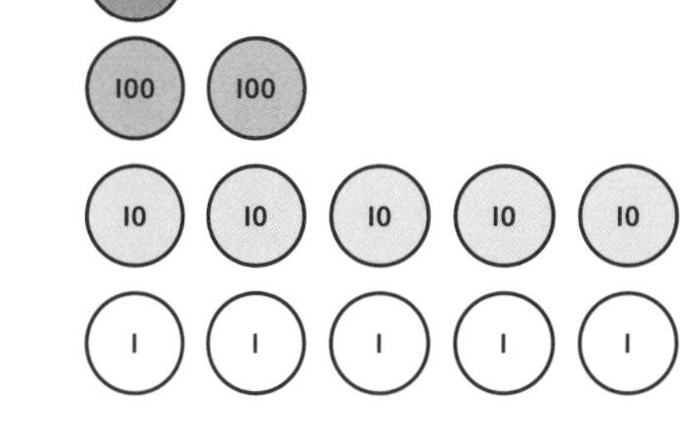

b)

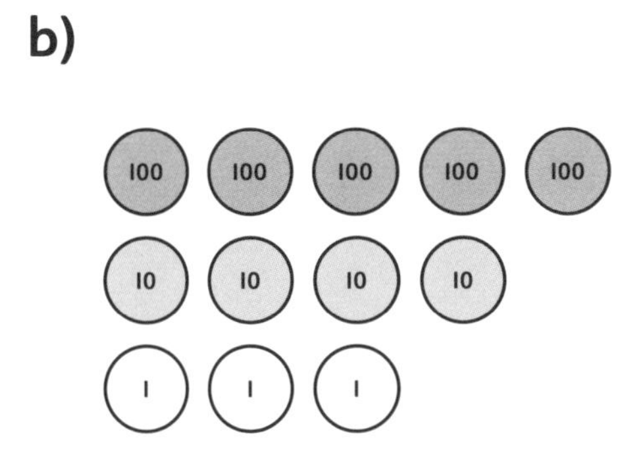

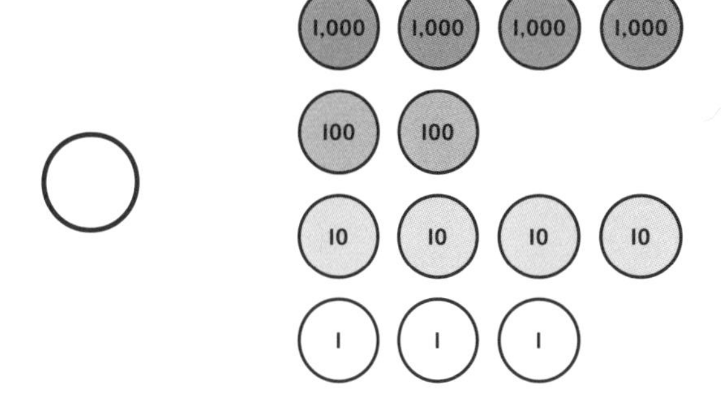

c)

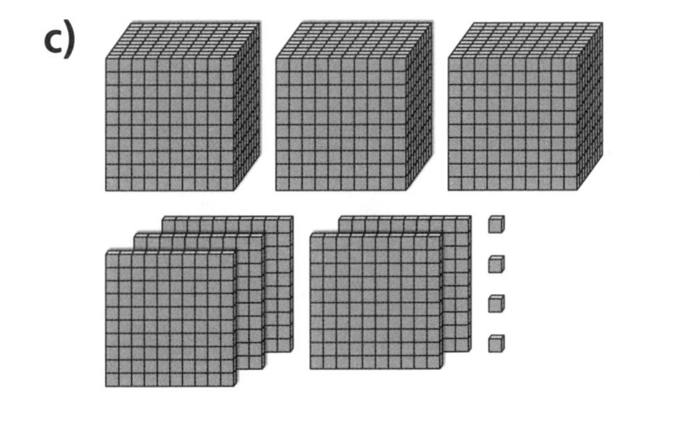

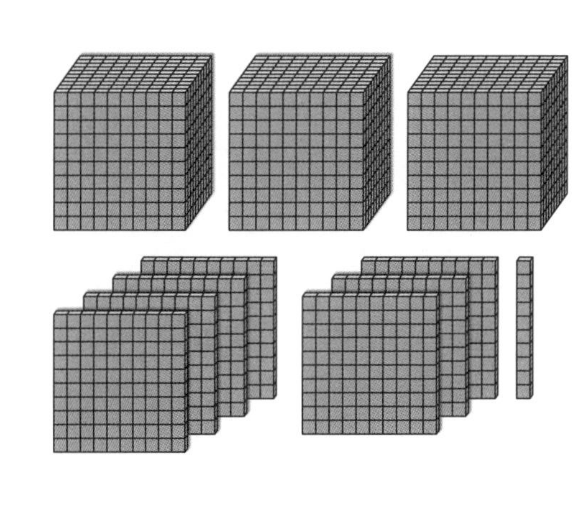

d)

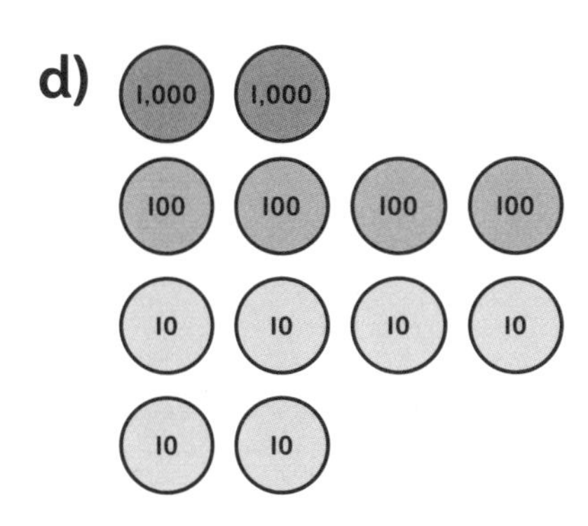

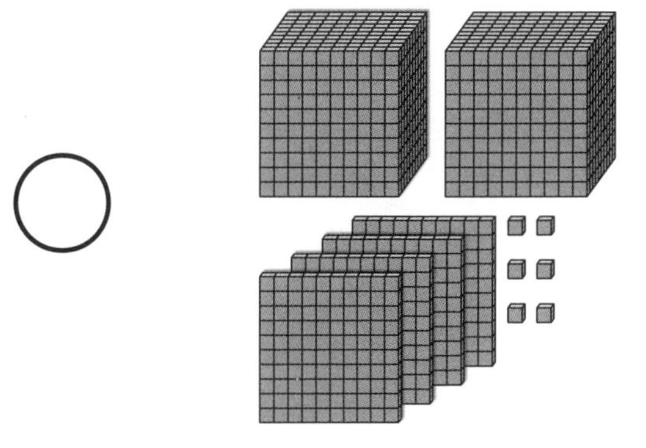

3 Draw counters in the boxes to complete the numbers.

Th	H	T	O

<

Th	H	T	O

Try to complete the boxes in more than one way.

Reflect

Compare these two numbers in different ways.

Th	H	T	O
1,000 1,000 1,000 1,000 1,000	100 100		1 1 1 1

Th	H	T	O

→ Textbook 4A p56

Comparing 4-digit numbers 2

1 Look at the numbers then complete the sentences.

a)

Th	H	T	O
5	0	1	5
4	3	0	1

☐ is smaller than ☐.

b)

Th	H	T	O
6	7	2	3
6	7	5	1

☐ is smaller than ☐.

c)

Th	H	T	O
	9	4	5
4	7	8	1

☐ is greater than ☐.

2 Complete the calculations using <, > or =

a) 3,560 ◯ 3,650

b) 8,034 ◯ 904

c) 5,000 + 1,200 + 30 + 4 ◯ 7,000 + 200 + 30 + 4

d) 9,451 ◯ 200 + 300 + 150 + 1

e) 5,760 ◯ 700 + 5,000 + 60

3 Complete the missing digits.

a) 4,☐78 < 4,592

b) 7,8☐9 > 7,8☐4

c) 5,04☐ < ☐,042

I wonder if there is more than one digit that would work in each empty box.

4 Amelia makes the following number:

Th	H	T	O

Tick all the numbers that are less than this number.

Th	H	T	O
2	5	0	0

☐

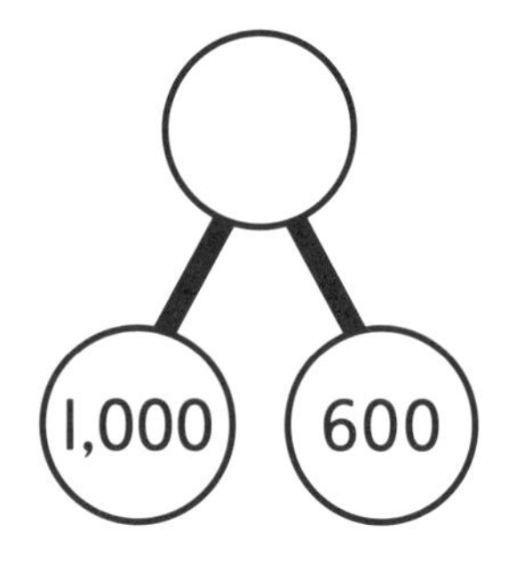

☐

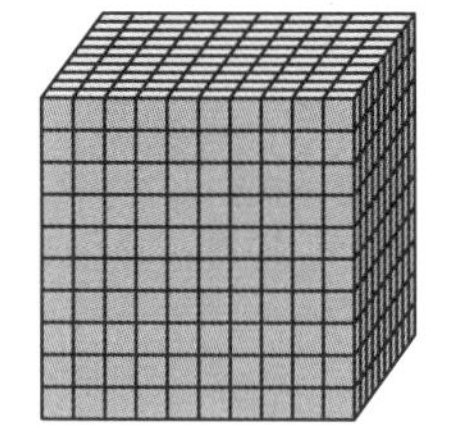
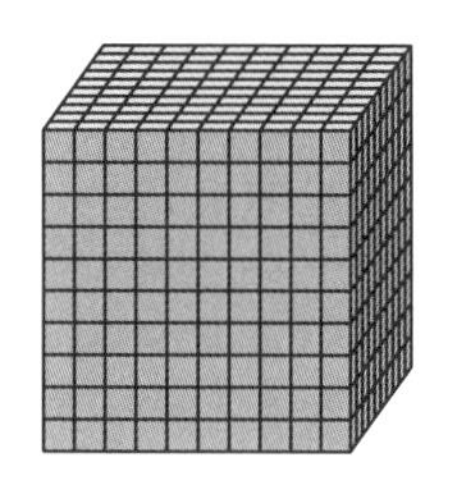
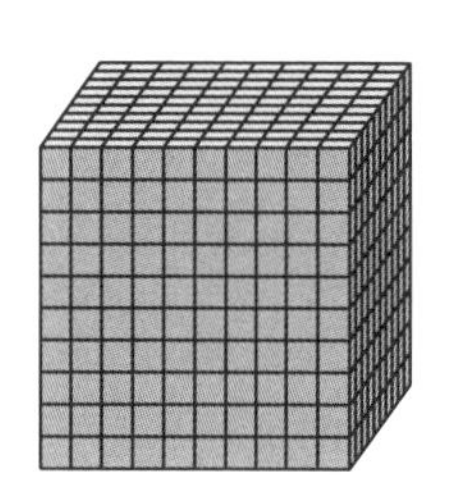

5 Here is a number line. 

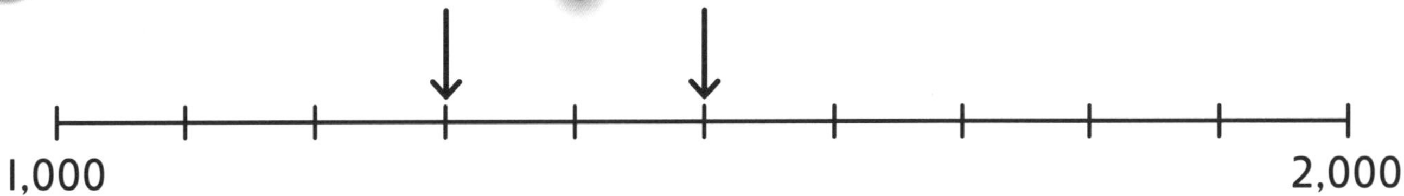

Write down 6 numbers that lie between the two arrows.

____

6 Use each of the digits from 1–6 once to complete the comparisons.

4,52☐ > 4,☐71

5,3☐6 < 5,☐74

☐,425 < 6,☐52

| 1 | 2 | 3 | 4 | 5 | 6 |

Try and complete the challenge in more than one way.

Reflect

highest, smallest, same, left, right, different

Choose the correct words to complete the sentences.

To compare two numbers, start with the ______________ place value column.

If the numbers are the ______________ , look at the column to the ______________ .

Continue until you find two numbers that are ______________ , so that you can compare.

→ Textbook 4A p60

Ordering numbers to 10,000

1 Order the numbers from largest to smallest.

Th	H	T	O
6	4	2	1
6	5	3	6
6	5	4	1
6	5	3	7

____ , ____ , ____ , ____

2 **a)** Write the numbers shown here.

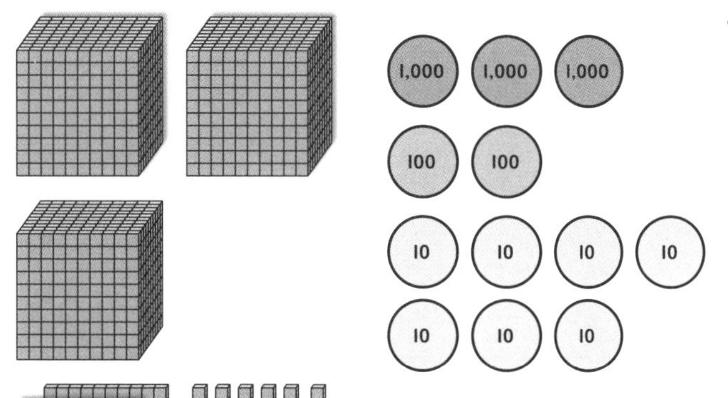

Three thousand, two hundred and fifty-eight

Th	H	T	O

____ ____ ____ ____

b) Write the numbers in ascending order.

____ ____ ____ ____

3 Order the numbers in descending order.

a) 4,502 kg, 3,821 kg, 4,314 kg, 4,099 kg

__________ , __________ , __________ , __________

b) 812 m, 8,032 m, 8,120 m, 7,830 m, 7,909 m

__________ , __________ , __________ , __________ , __________

4 Max, Lexi and Richard track their activities for a month to see how far they swim, run and cycle.

	Swim	Run	Cycle
Max	2,500 m	3,400 m	7,850 m
Lexi	750 m	4,500 m	7,995 m
Richard	2,350 m	4,180 m	7,855 m

a) Who swam the furthest? __________

b) Who ran the 2nd shortest distance? __________

c) Put the distances they cycled in order starting with the shortest.

__________ , __________ , __________

5 Fill in the missing digits.

a) 3,246 < 3,[ ]48 < 3,312 < 3,3[ ]1

b) 5,[ ]74 > 5,6[ ]2 > 5,66[ ] > 5,663

c) 2,710 < [ ] < [ ] < [ ] < 2,900

CHALLENGE

6 Zac has put five 4-digit numbers in ascending order.

4,317, __________, __________, __________, 4,353

All the numbers have a digit total of 15.

What are the other three numbers?

A digit total is the sum of all the digits in a number.

Reflect

Use the digits 5, 6, 8 and 9 to make some 4-digit numbers.

Then write your numbers in descending order.

__________, __________, __________, __________, __________,

__________, __________, __________, __________, __________,

__________, __________

→ Textbook 4A p64

Rounding to the nearest 1,000

1 Round each number to the nearest 1,000.

a)

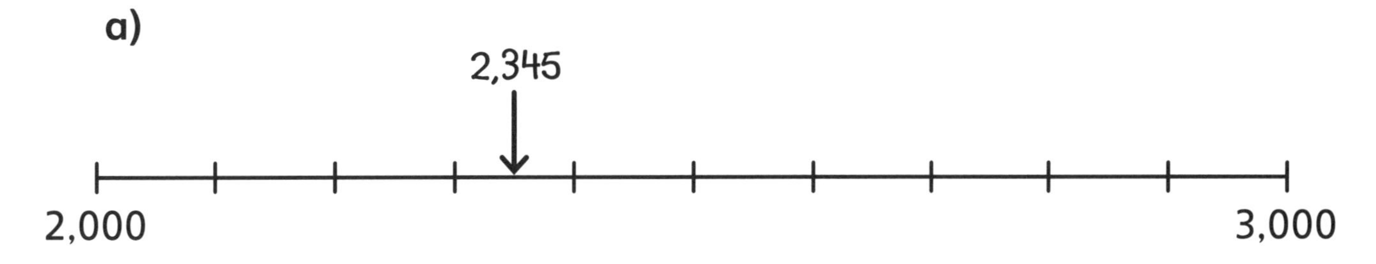

2,345 rounded to the nearest 1,000 is ☐.

b)

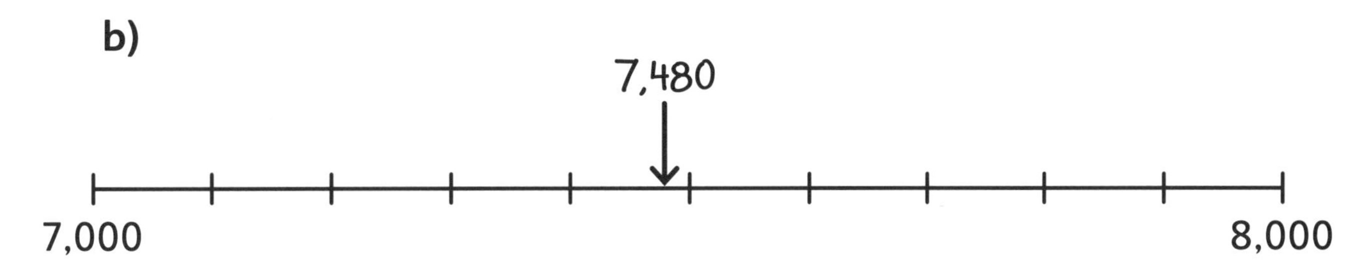

7,480 rounded to the nearest 1,000 is ☐.

c)

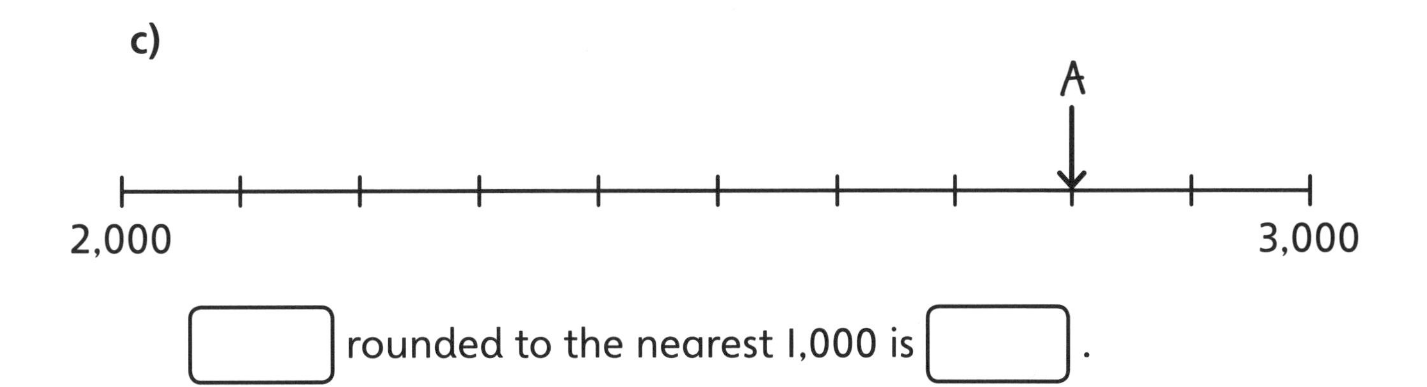

☐ rounded to the nearest 1,000 is ☐.

2 On the number line, mark five numbers that round to 5,000 to the nearest 1,000.

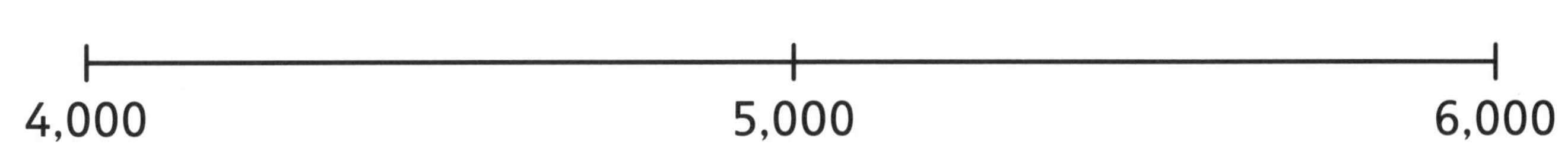

3 Round each of these numbers to the nearest 1,000.

a)

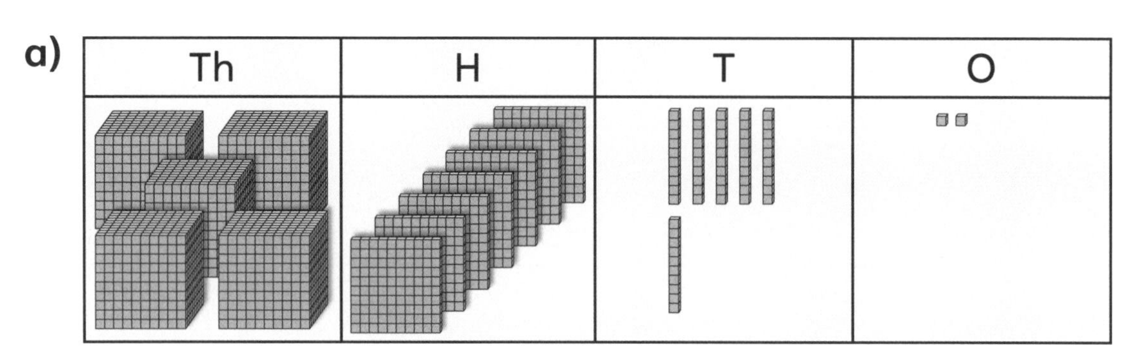

[ ] rounded to the nearest 1,000 is [ ].

b)

Th	H	T	O
1,000 1,000	100 100 100	10 10 10 10 10 10 10 10	

[ ] rounded to the nearest 1,000 is [ ].

4 Complete the following sentences.

a) 2,500 to the nearest 1,000 is [ ].

b) 3,180 to the nearest 100 is [ ].

c) 5,050 to the nearest 100 is [ ].

5 Complete the following sentences.

a) 4,997 to the nearest 1,000 is [ ].

b) 4,997 to the nearest 100 is [ ].

c) 4,997 to the nearest 10 is [ ].

6 When a number is rounded to the nearest 1,000, it is 9,000.

CHALLENGE

When the same number is rounded to the nearest 100, it is 8,900.

When the same number is rounded to the nearest 10, it is 8,910.

What could the number be? List all the possibilities.

____

____

Reflect

Use the digits 5, 2, 7 and 0 to make some 4-digit numbers.

Make a number that rounds to 5,000. Make a number that rounds to 2,000.

Is there more than one answer for each? Explain to a friend the method that you used.

____

____

→ Textbook 4A p68

Solving problems using rounding

1 The table below shows how many people took part in a fun run in three different cities.

Round each number to the nearest 100.

City	Number of runners	To the nearest 100
Manchester	8,498	
Leeds	7,849	
Birmingham	8,805	

2 Luis and Emma are comparing how far they have cycled this week.

Who has cycled the furthest, to the nearest 1,000 metres?

____

3 A farmer counts his crops from the harvest. 

Round each crop to the nearest 1,000.

Crop	Number	To the nearest 1,000
Potatoes	9,451	
Carrots	9,050	
Parsnips	5,500	
Turnips	3,900	

4 Andy says, 'It does not matter whether I round to the nearest 10, 100 or 1,000, I get the same number.'

Is Andy correct? Explain your answer.

____

____

5 Add the missing digits in the number column. Then round the numbers.

Number	Rounded to the nearest 1,000	Rounded to the nearest 100	Rounded to the nearest 10
8,341			
6,☐☐2			6,890
8,☐7☐	9,000		
5,☐5☐		5,500	
☐,☐☐7		6,100	6,100

6 Bella uses 8 counters to make a 4-digit number.
The number rounds to 4,000 to the nearest 1,000.

Th	H	T	O

How many different ways can Bella make a 4-digit number that has at least one counter in each column?

____

What is the greatest number she can make?

____

What is the smallest number she can make?

____

Reflect

Aki's number rounded to the nearest 10, 100 and 1,000 is 2,000.
What could Aki's number be?

Could there be more than one answer to this question?

____

____

____

____

→ Textbook 4A p72

Counting in 25s

1 Books are packed in boxes of 25s.

Work out how many books are stacked in each pile below.
Write your answers in the table.

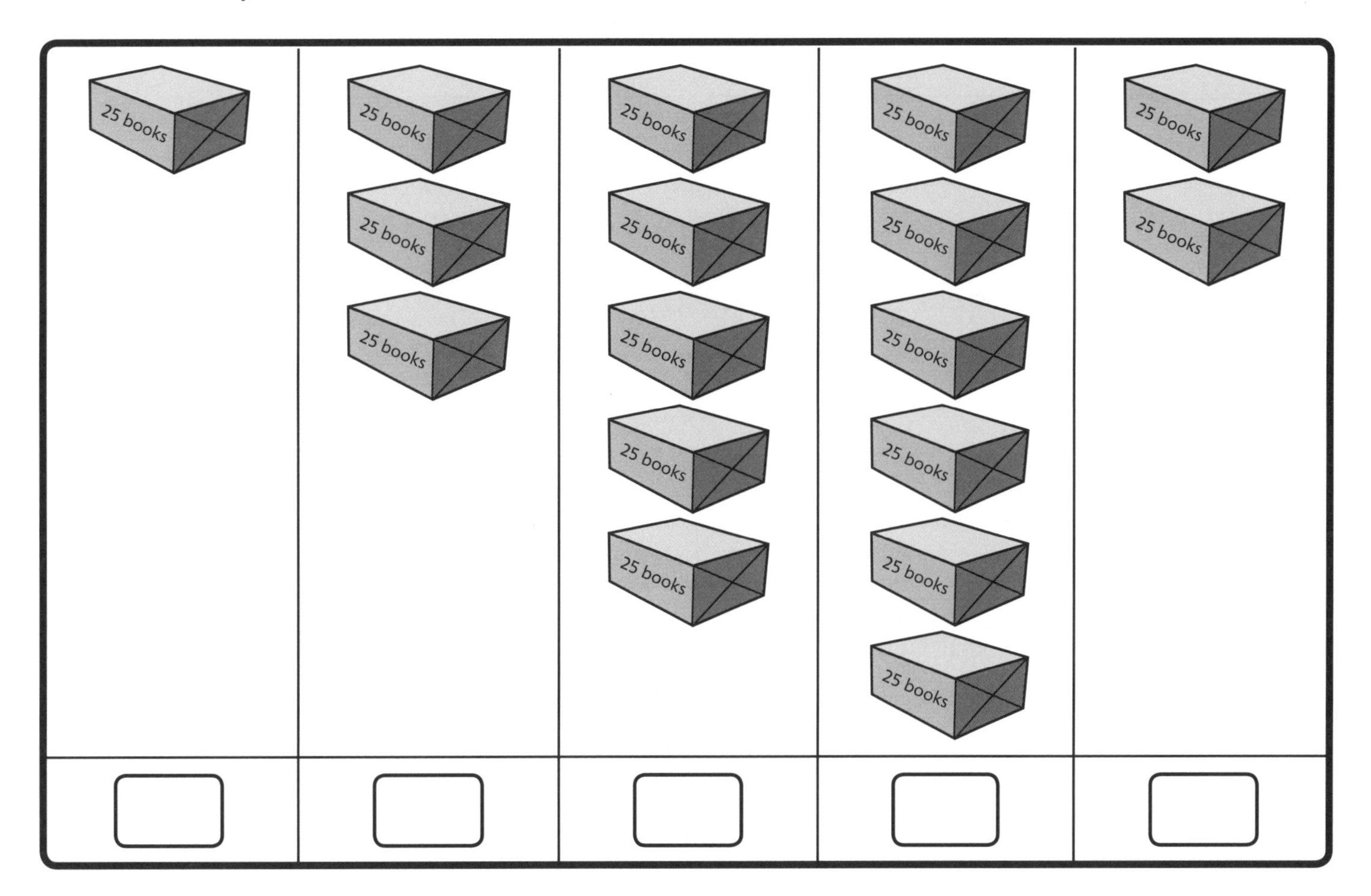

2 Complete each number line.

a)

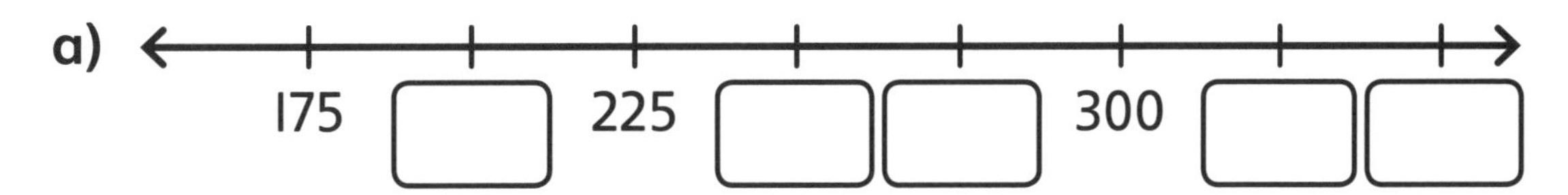

b)

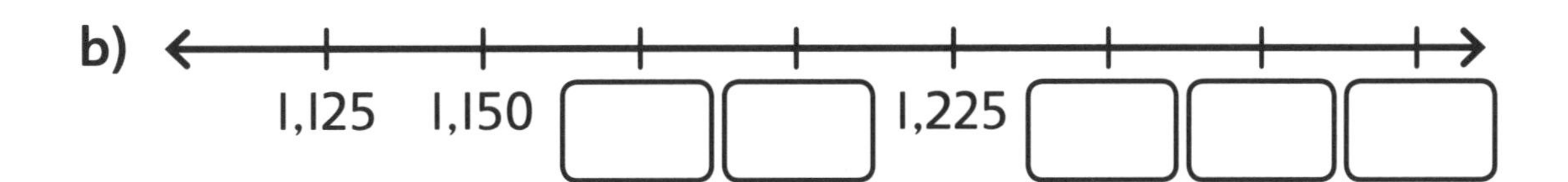

3 Fill in the gaps below.

a)

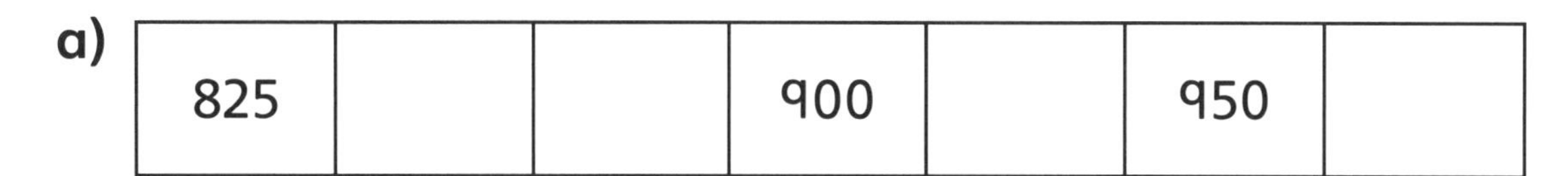

825			900		950	

b)

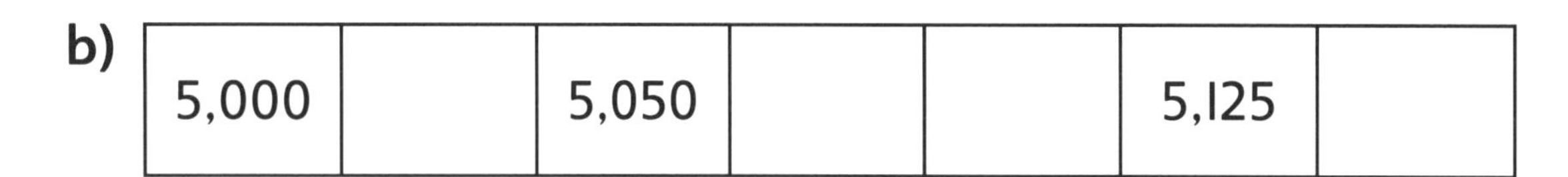

5,000		5,050			5,125	

4 Richard is counting up in 25s from 0.

Circle the numbers that he will say.

165 155 200 205

250 325 405 475

502 900 975 1,000

How can you tell which numbers Richard will say? Explain your answer.

____

____

5 Mo plays a video game. For each alien spaceship he destroys, he gets 25 points.

He scores 325 points. How many alien spaceships did Mo destroy?

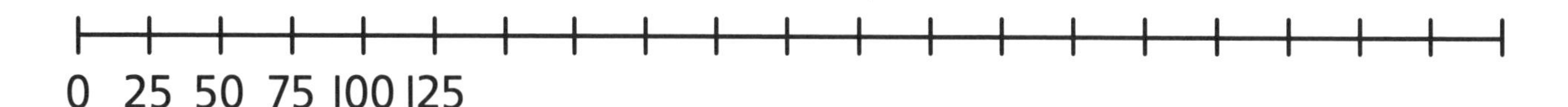

Mo destroyed ☐ alien spaceships.

CHALLENGE

6 Stickers come in packs of 25.

How many packets will Luis need to buy if he wants 313 stickers?

[ ]

How many packets will Reena need to buy if she wants 999 stickers?

[ ]

25 stickers

Reflect

Olivia is counting backwards in 25s from 5,025 to 0.

Circle the numbers she will say.

5,000	4,010	4,075	0	100	1,070
2,005	95	1,000	99	50	52

Explain your method.

→ Textbook 4A p76

Negative numbers 1

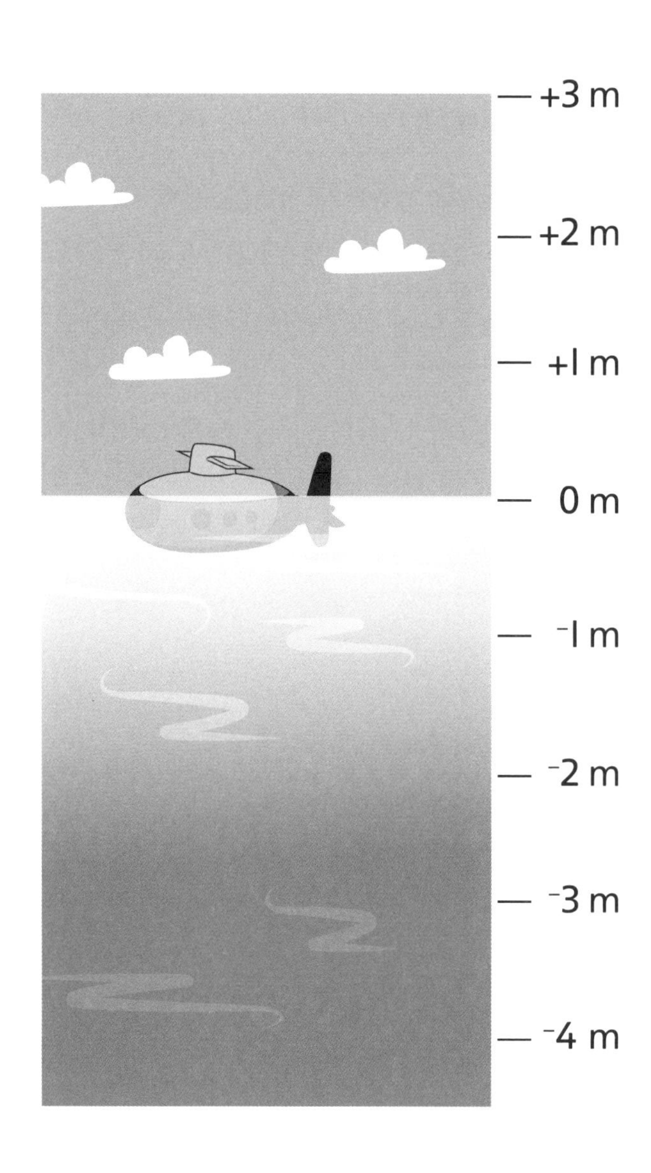

1 **a)** A submarine is on the surface of the sea.

It goes down 3 metres.

What level is it at now?

The submarine is at ☐ metres.

b) The submarine is at $^{-}2$ metres.

A crane lifts it 5 metres up.

What level is it at now?

The submarine is at ☐ metres.

2

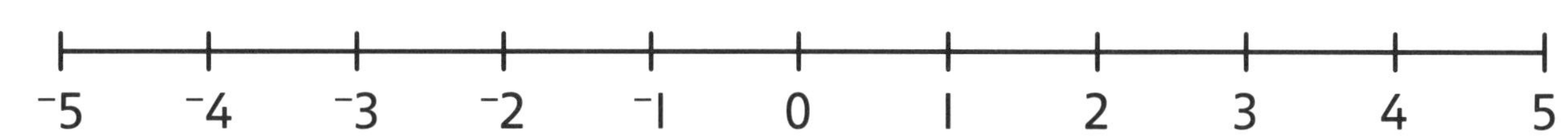

a) Starting at $^{-}4$, count forwards 5.

The finish number is ☐.

b) Starting at 3, count backwards 7.

The finish number is ☐.

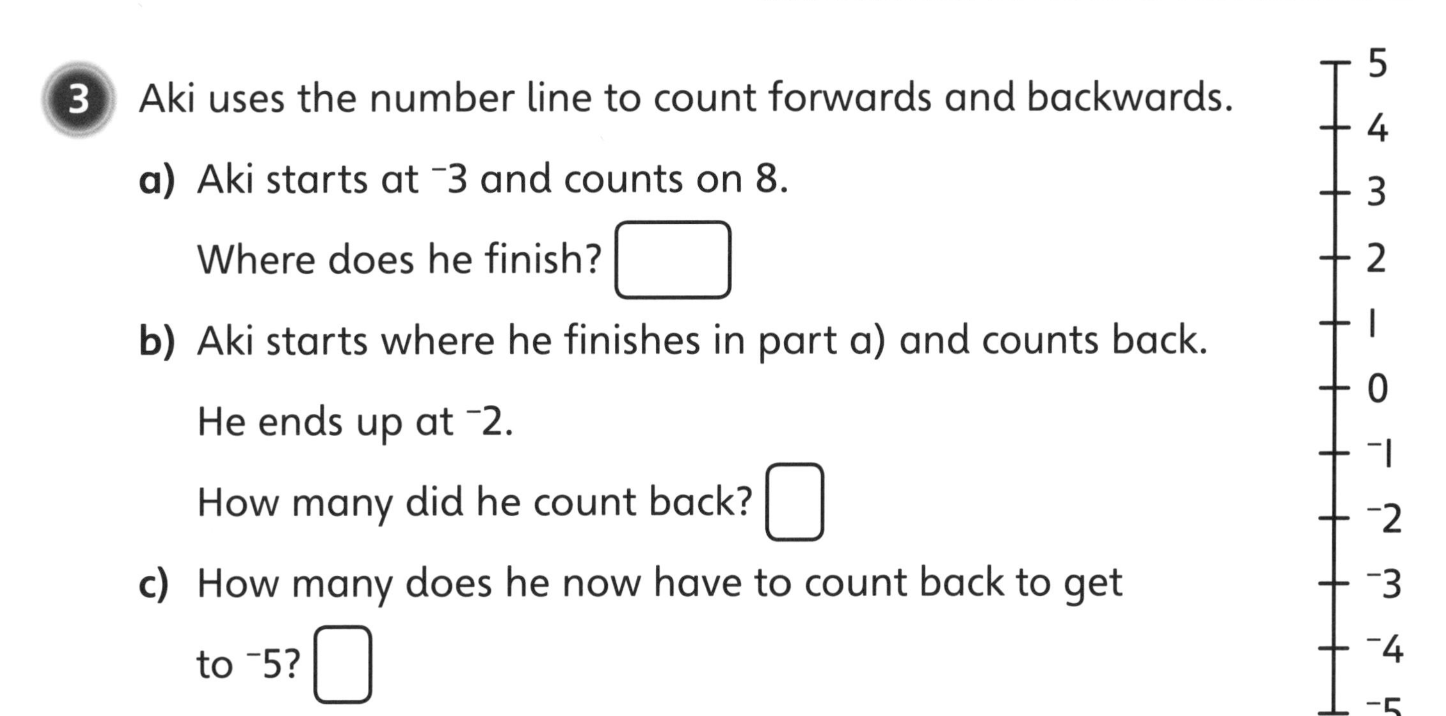

3 Aki uses the number line to count forwards and backwards.

a) Aki starts at $^{-}3$ and counts on 8.

Where does he finish? ☐

b) Aki starts where he finishes in part a) and counts back.

He ends up at $^{-}2$.

How many did he count back? ☐

c) How many does he now have to count back to get to $^{-}5$? ☐

4 A board game looks like this:

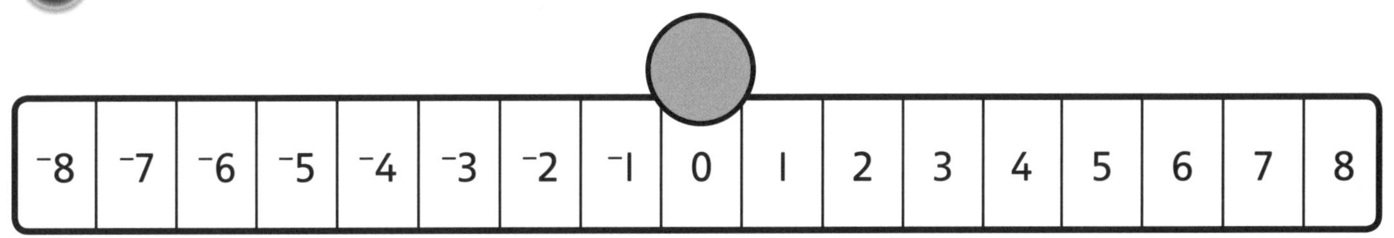

Player 1 starts on 0, then moves forwards 7, then backwards 3, then forwards 2, then backwards 8, then backwards 3. Where does the player finish?

☐

5 Write the next three numbers in the sequences.

a) 3, 2, 1, 0, ☐, ☐, ☐

b) $^{-}7$, $^{-}6$, $^{-}5$, ☐, ☐, ☐

c) 6, 4, 2, ☐, ☐, ☐

d) $^{-}12$, $^{-}9$, $^{-}6$, ☐, ☐, ☐

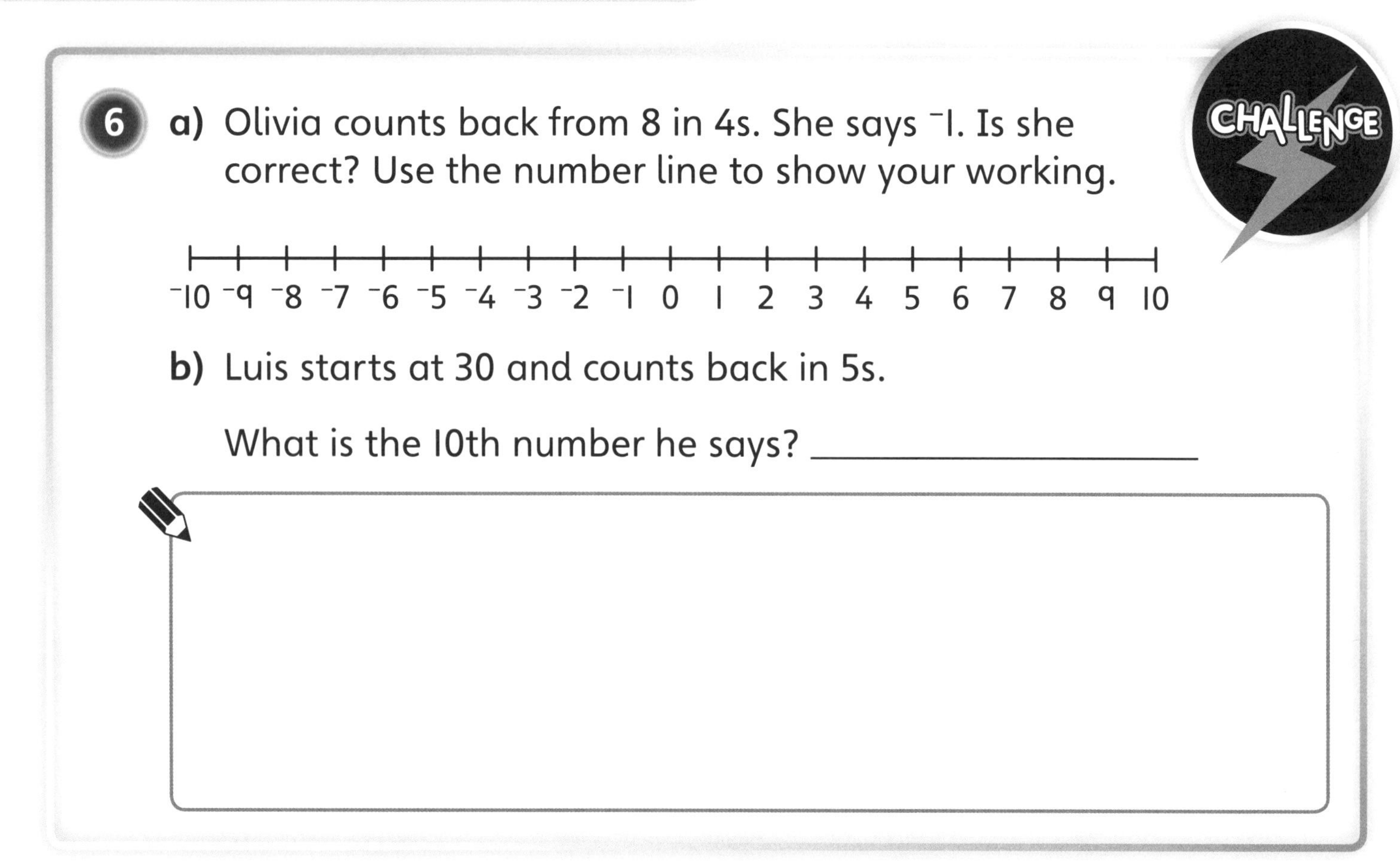

6 **a)** Olivia counts back from 8 in 4s. She says ⁻1. Is she correct? Use the number line to show your working.

b) Luis starts at 30 and counts back in 5s.

What is the 10th number he says? ________________

Reflect

⁻8	⁻7	⁻6	⁻5	⁻4	⁻3	⁻2	⁻1	0	1	2	3	4	5	6	7	8

Start at 0. Roll a dice and move forwards that number. Count as you move up the number track. Roll the dice again and move backwards that number.

Repeat several times, moving forwards then backwards.

Explain how you could predict where you will land without counting.

→ Textbook 4A p80

Negative numbers 2

1 Complete the number lines.

a)

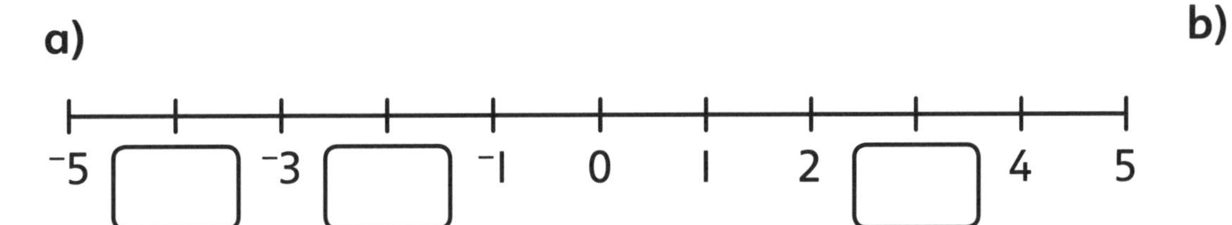

b)

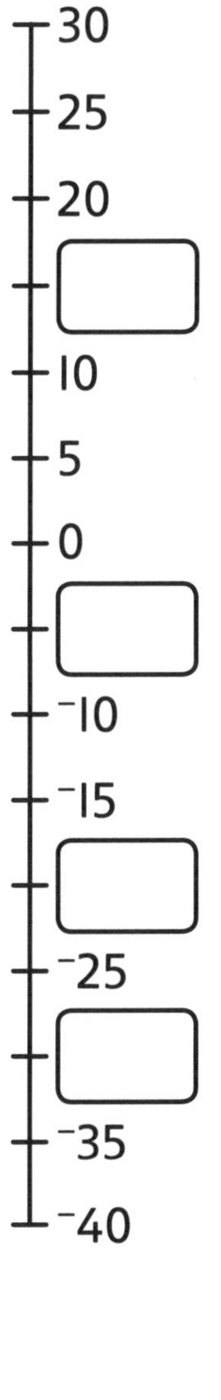

2 Complete the thermometers and write the temperature.

a) The temperature is ☐ degrees Celsius.

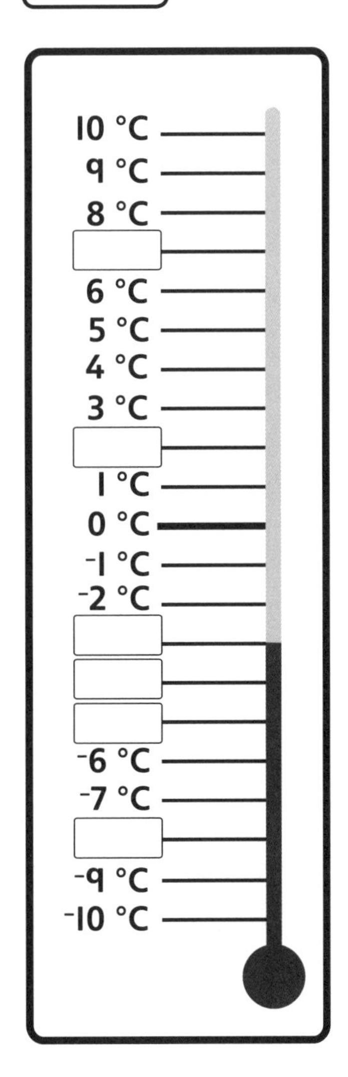

b) The temperature is ☐ degrees Celsius.

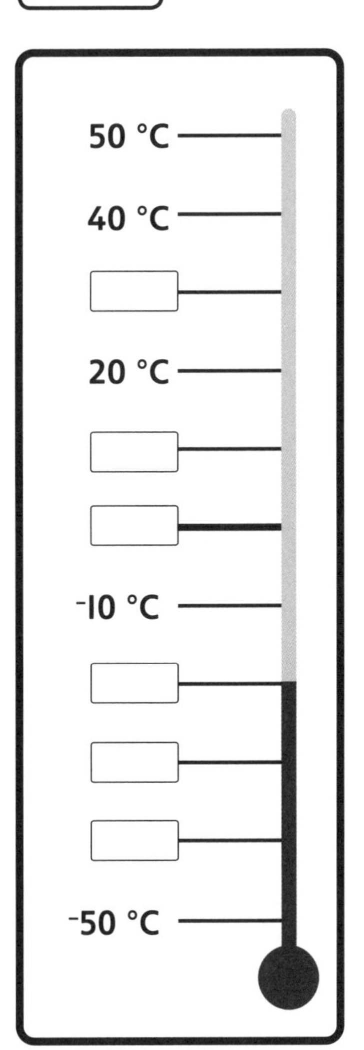

3 Complete the missing numbers.

a)

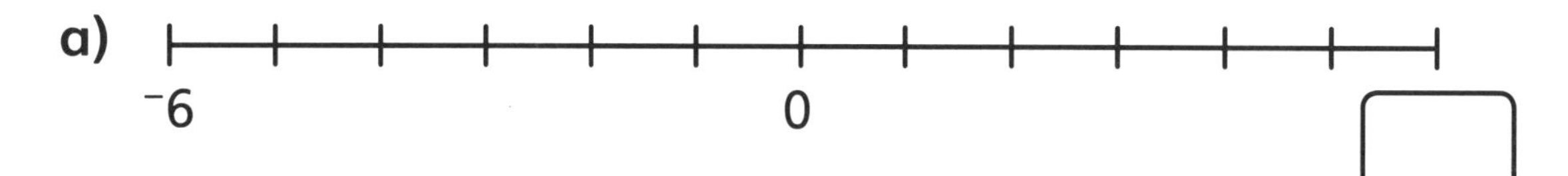

b)

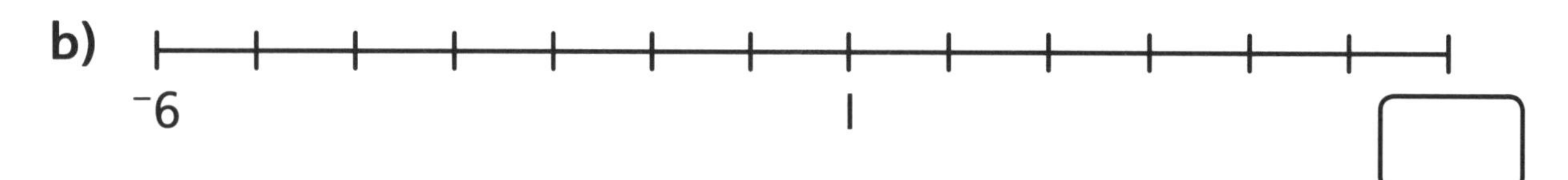

c)

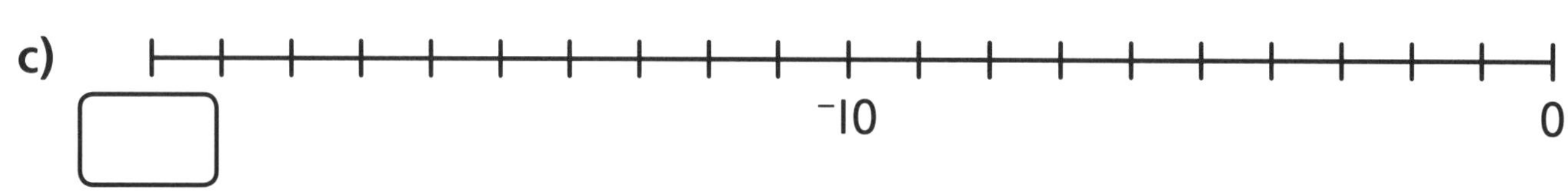

4 Write the new temperatures after the temperature change.

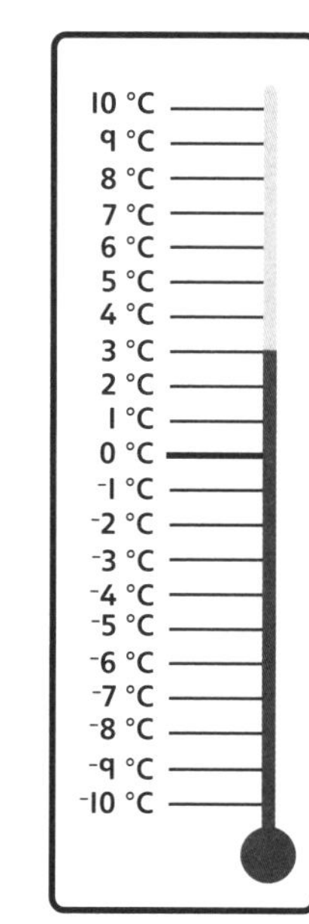

a) The temperature drops by 5 degrees Celsius.

The temperature is now ☐ degrees Celsius.

b) The temperature then goes up by 7 degrees Celsius.

The temperature is now ☐ degrees Celsius.

5 Explain the mistakes on each of these number lines.

a)

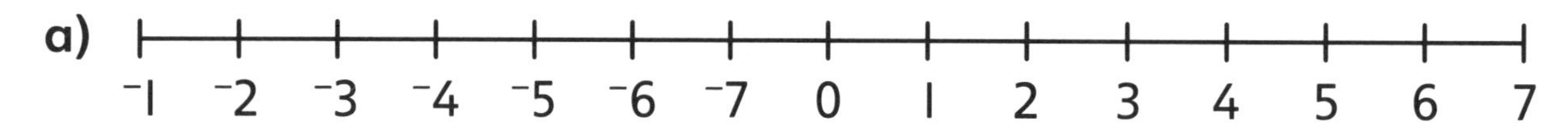

____

____

b)

7 6 5 4 3 2 1 0 $^{-}1$ $^{-}2$ $^{-}3$ $^{-}4$ $^{-}5$ $^{-}6$ $^{-}7$

____

____

6 Estimate the position of the letters on the number line.

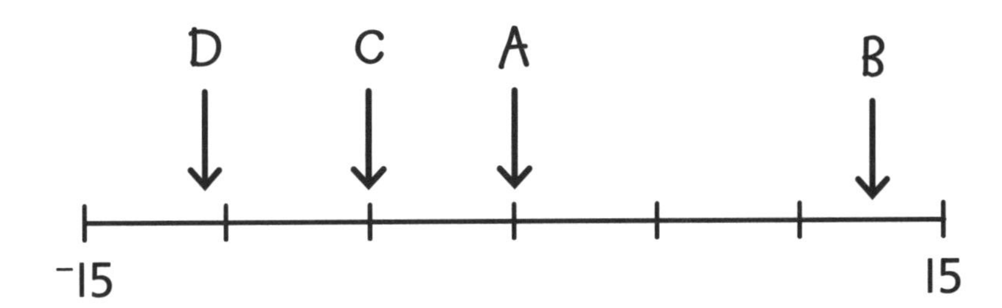

A = ☐ B = ☐ C = ☐ D = ☐

7 Complete the number line in two different ways.

☐ ☐ ⁻3 ☐ ☐ ☐ ☐

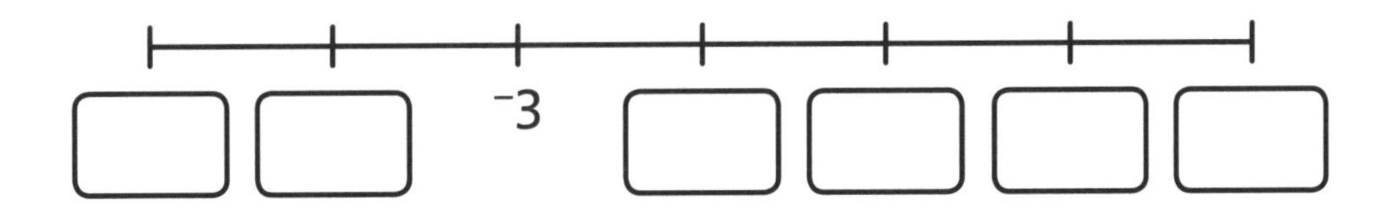

☐ ☐ ⁻3 ☐ ☐ ☐ ☐

Reflect

Max says, '⁻1 is less than ⁻4 because 1 is smaller than 4.'

Do you agree with Max? Explain your answer.

____

____

→ Textbook 4A p84

End of unit check

My journal

1 When finding 1,000 less than a number, which place value columns will never change? Use the grid to help you.

Th	H	T	O

When rounding a number to the nearest 1,000, which place value columns can change?

Write an example.

Power check

How do you feel about your work in this unit?

Power play

What you need:

A blank place value grid each.

Th	H	T	O

A 0–9 dice and a 0–6 dice to share between your pair.

Six number cards as follows:

1	2	3	4	5	6
Round to the nearest 1,000	Round to the nearest 100	Round to the nearest 10	What is 1,000 more than this number?	What is 1,000 less than this number?	What is 100 more than this number?

Roll the 0–9 dice four times each.

After each roll, write the number on the dice in one of the place value columns on your grid until you each have a number in every column.

Lay the number cards out in front of you. Roll the 0–6 dice once each. Choose the number card that matches the number on the dice that you rolled.

Do what the number card asks to the number on your grid. Score a point for each correct answer. Roll the 0–6 dice again and try a different number card.

Play again. This time select the card first, then roll the dice four times. A point is scored for the greatest possible answer each time.

→ Textbook 4A p88

Adding and subtracting 1s, 10s, 100s, 1,000s

1 Solve these calculations.

a)

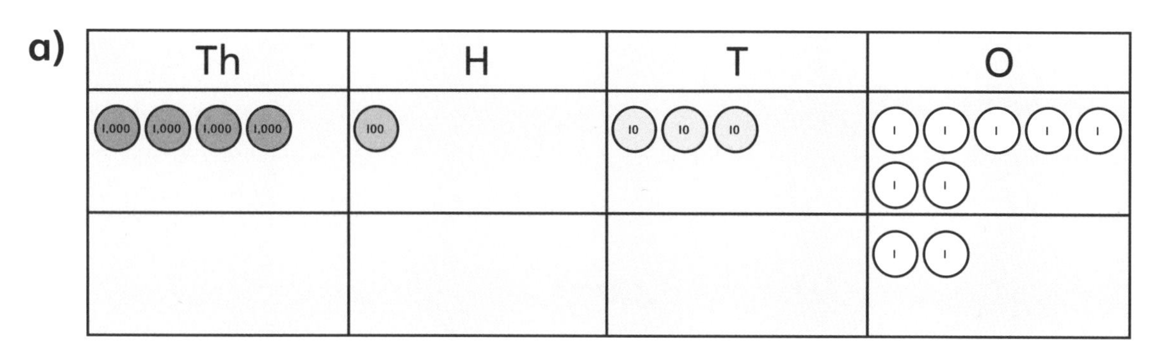

4,137 + 2 = ☐

b)

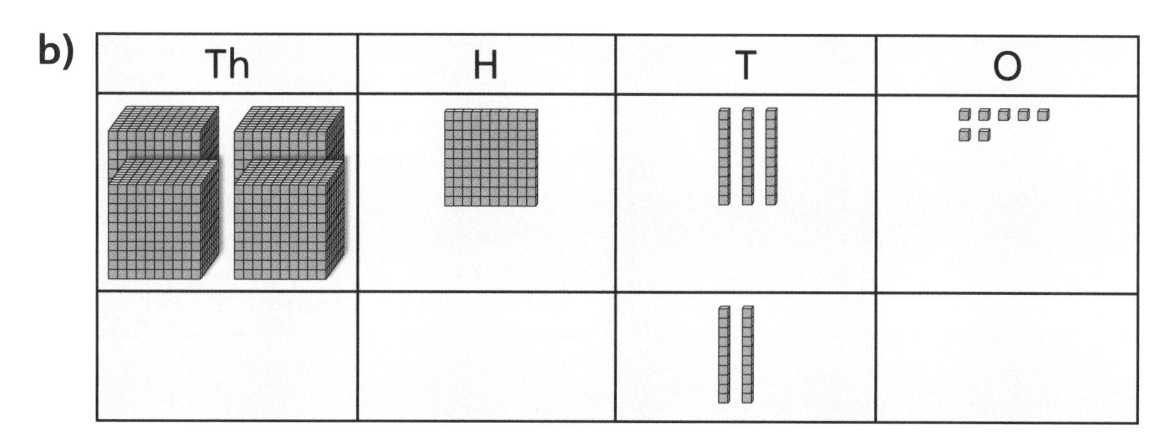

4,137 + ☐ = ☐

2 Work out the missing numbers.

Th	H	T	O
1,000 1,000 1,000 1,000 1,000 1,000	100 100 100 100 100 100	10 10 10 10 10 10	1 1 1 1 1 1

a) 6,666 + 2 = ☐

b) 6,666 + 20 = ☐

c) 2,000 + 6,666 = ☐

d) 6,666 – 200 = ☐

e) 6,666 = ☐ – 200

3 Complete these calculations.

a) 3,154 + 500 = ☐

b) 500 + 4,351 = ☐

c) 9,786 – 4,000 = ☐

d) ☐ = 7,968 – 400

e) ☐ + 1,000 = 2,134

f) ☐ + 4,000 = 4,521

g) 4,014 – 10 = ☐

h) 5,001 – ☐ = 1

4 
a) How much does the car cost now?

7,999 ◯ ☐ = ☐

The car costs £☐ now.

b) How much has the price of the van changed by?

8,749 ◯ ☐ = 8,249.

The price has changed by £☐.

5 3,333 + 4,000 = 7,333

Explain how to use this fact to solve 7,333 – 3,333 = .

Now work out 8,181 – 8,111 = ☐

6 **a)** Use these cards once each to complete all the puzzles.

100	200	300	400	500	600	700	800	900

3,334 + ☐ – ☐ = 3,434   3,334 – ☐ + ☐ = 3,434

3,934 – ☐ – ☐ = 3,434

3,434 – ☐ – ☐ + ☐ = 3,434

b) Find another way to do them.

3,334 + ☐ – ☐ = 3,434   3,334 – ☐ + ☐ = 3,434

3,934 – ☐ – ☐ = 3,434

3,434 – ☐ – ☐ + ☐ = 3,434

Reflect

5,167 + ☐ = 9,167

Show how to work out the missing number.

→ Textbook 4A p92

Adding two 4-digit numbers 

1 Holly has saved £2,321. Toshi has saved £525.

How much have they saved altogether?

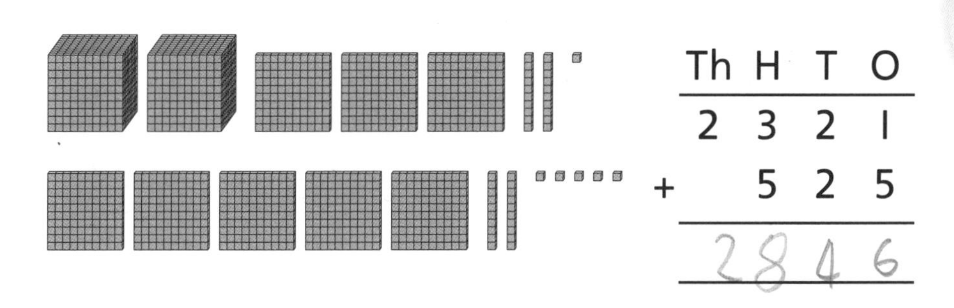

	Th	H	T	O
	2	3	2	1
+		5	2	5
	2	8	4	6

They have saved £ 2846 altogether.

2 Complete the additions.

a)

Th	H	T	O
1,000 1,000 1,000	100		1 1 1 1 1
1,000 1,000 1,000	100 100 100 100 100	10	1

	Th	H	T	O
	3	1	0	5
+	3	5	1	1
	6	6	1	6

3,105 + 3,511 = 6616

b)

Th	H	T	O
1,000 1,000 1,000 1,000 1,000	100	10 10 10	1

	Th	H	T	O
	5	1	3	1
+	3	0	5	1
	8	1	8	2

5,131 + 3,051 = 8182

3 Fill in the missing digits.

a)

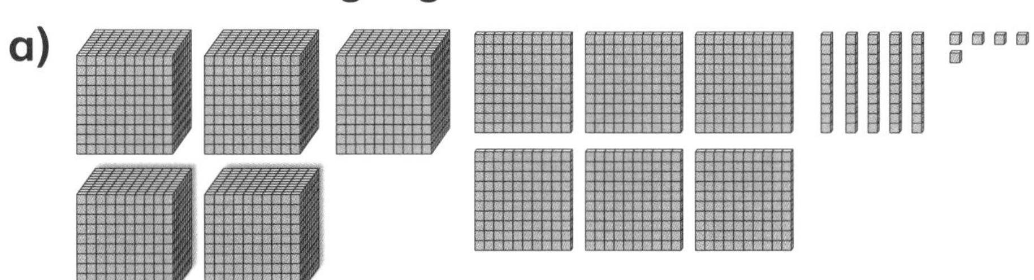

	Th	H	T	O
	4	5	1	3
+				
	5	6	5	6

b)

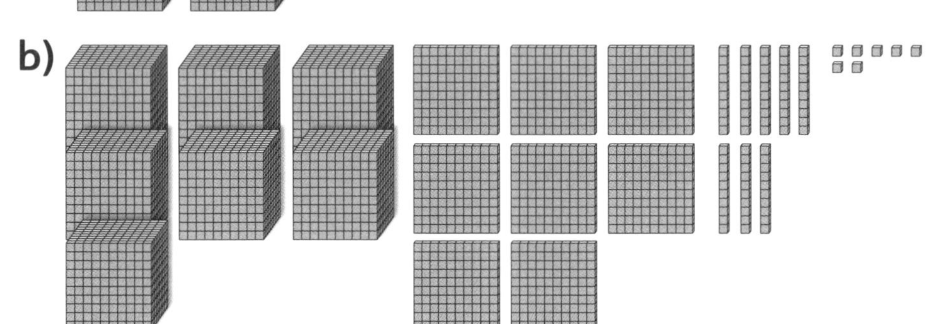

	Th	H	T	O
	3		7	
+		8		6
	7	8	8	7

4 Find and correct the two mistakes.

a) 3,452 + 42 = ☐

	Th	H	T	O
	3	4	5	2
+	4	2		
	7	6	5	2

	Th	H	T	O
+				

b) 1,025 + 1,500 = ☐

	Th	H	T	O
	1	2	0	5
+	1	5	0	0
	2	7	0	5

	Th	H	T	O
+				

5 What is 2,345 more than 4,153?

6 Complete these calculations. Show which method you used for each.

a) 1,045 + 2,331 = ☐ **b)** 4,521 + 432 = ☐

7 How many different solutions can you find using only the digits 1 and 8?

Can you find all the possible solutions?

	Th	H	T	O
+				
	9	9	9	9

I think there are ☐ different solutions because …

Reflect

Work out 2,512 + 5,105 using column addition. Choose some equipment to show a partner how you did it.

→ Textbook 4A p96

Adding two 4-digit numbers ❷

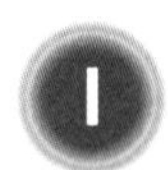

a) Ebo ran 1,175 m. Lee ran 1,750 m. How far did they run in total?

Th	H	T	O
1,000	100	10 10 10 10 10 10 10	1 1 1 1 1
1,000	100 100 100 100 100 100 100	10 10 10 10 10	
	100		

	Th	H	T	O
	1	1	7	5
+	1	7	5	0

☐ + ☐ = ☐

They ran ☐ m in total.

b) Kate ran 2,400 m and Bella ran 975 m further than Kate. How far did Bella run?

Th	H	T	O
	100 100 100 100 100 100 100 100 100	10 10 10 10 10 10 10	1 1 1 1 1
1,000 1,000	100 100 100 100		

	Th	H	T	O
		9	7	5
+	2	4	0	0

☐ + ☐ = ☐

Bella ran ☐ m.

c) Lexi and Luis both ran 1,245 m. How far did they run altogether?

They ran ☐ m in total.

	Th	H	T	O
+				

2 Solve these additions using the column method.

a) 1,475 + 3,711 =

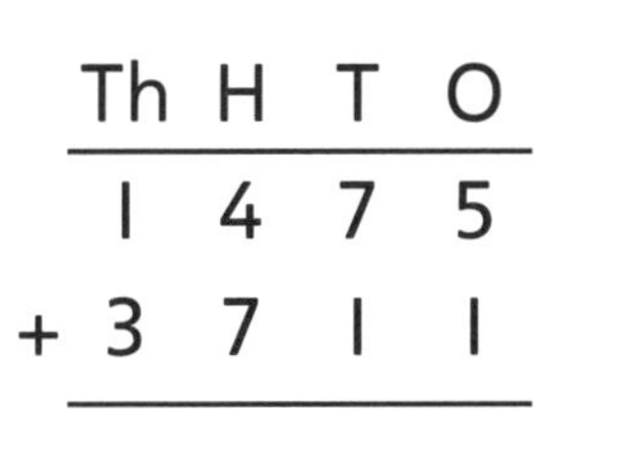

c) = 1,054 + 5,094

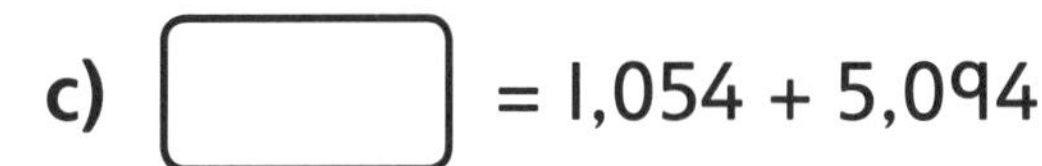

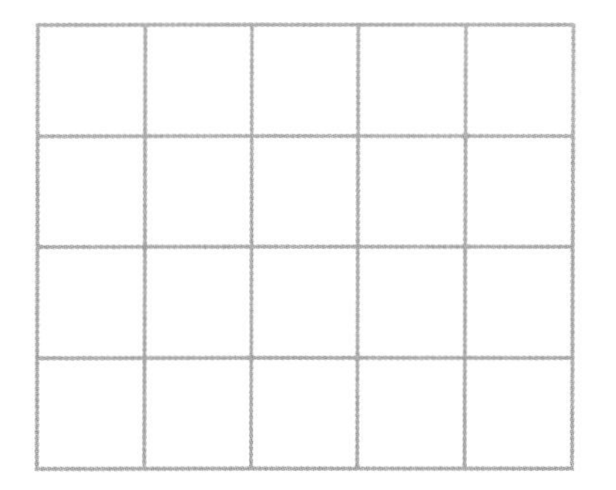

b) = 3,029 + 2,963

d) 179 + 2,608 =

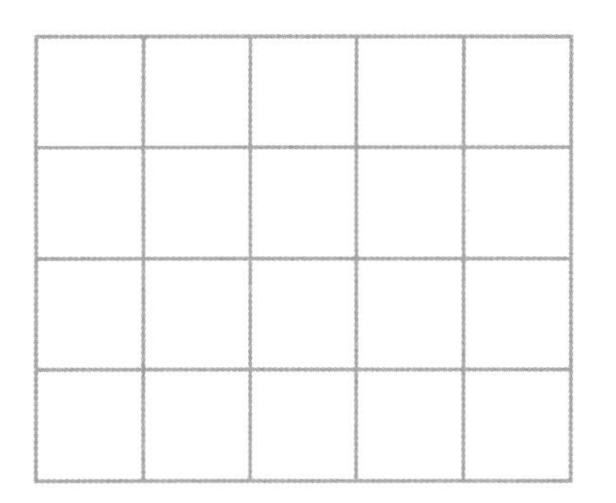

3 Complete each story problem so that it only has an exchange of 10s, and then show the number sentence to solve the problem.

a) There were 1,259 adult tickets sold and ☐ children's tickets sold. How many ______________________________?

b) There were ☐ seats on the left side and ☐ seats on the right side. How many ______________________________?

4 Find the missing digits. 

a)

	Th	H	T	O
	1	1	1	1
+				
	2	2	5	0

b)

	Th	H	T	O
+	1	8	2	3
	3	4	5	6

5 **a)** Solve ☐ = 1,575 + 5,520

	Th	H	T	O
+				

b) Now use that addition to solve these:

4,520 + 1,575 = ☐

☐ = 5,519 + 1,576

☐ = 1,565 + 5,510

☐ = 575 + 520

Reflect

Create three different additions that have one exchange of:

a) 1s

	Th	H	T	O
+				

b) 10s

	Th	H	T	O
+				

c) 100s

	Th	H	T	O
+				

→ Textbook 4A p100

Adding two 4-digit numbers 3

1 Complete these additions.

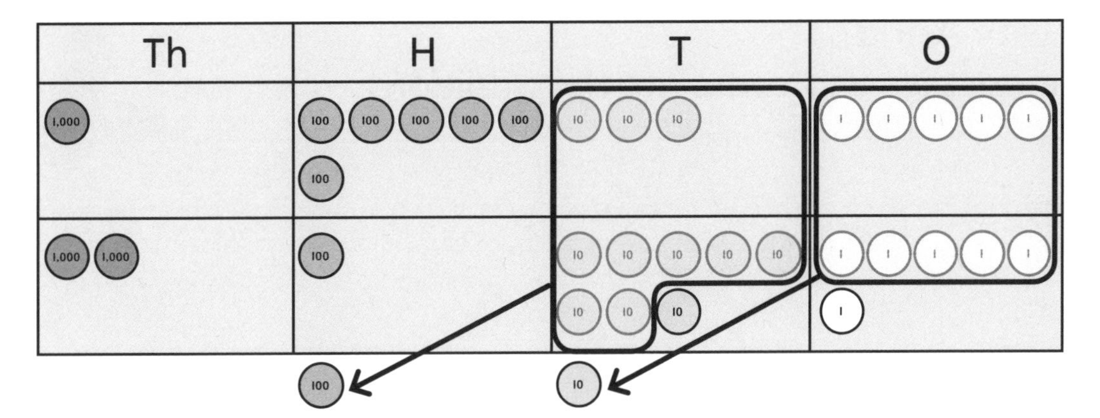

	Th	H	T	O
		1	6	3	5
+	2	1	8	6

Th	H	T	O
1,000 1,000	100 100 100 100	10 10 10 10 10 10	1 1 1 1 1
1,000	100 100 100 100 100 100	10 10 10 10 10 10	1 1

	Th	H	T	O
	2	4	6	5
+	1	6	6	2

2 **a)** Choose pairs of numbers so that each addition has two exchanges. Then solve each of your calculations.

3,405 1,726

1,283 199

	Th	H	T	O
+				

	Th	H	T	O
+				

b) Now think of your own numbers to make up two more additions, each with two exchanges.

	Th	H	T	O
+				

	Th	H	T	O
+				

3 [ ] = 1,218 + 3,783

I think this will only need one exchange. Only the 1s digits add to more than 9.

Do you agree with Astrid?
Explain to your partner and complete the addition.

4 Solve these additions.

a) 1,257 + 189 = [ ]

b) [ ] = 1,011 + 989

5 Complete these additions. Show your method.

I can see a mental method.

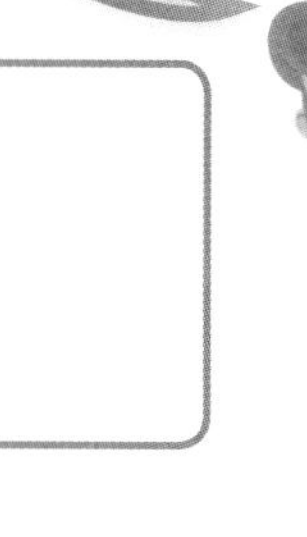

a) 654 + 2,999 = [ ]

b) 4,999 + 2,999 = [ ]

6 **a)** Fill in the missing digits in these calculations.

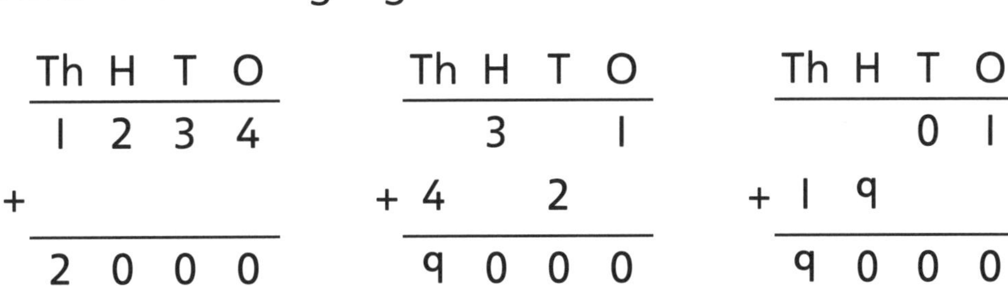

b) Find the size of each jump below.

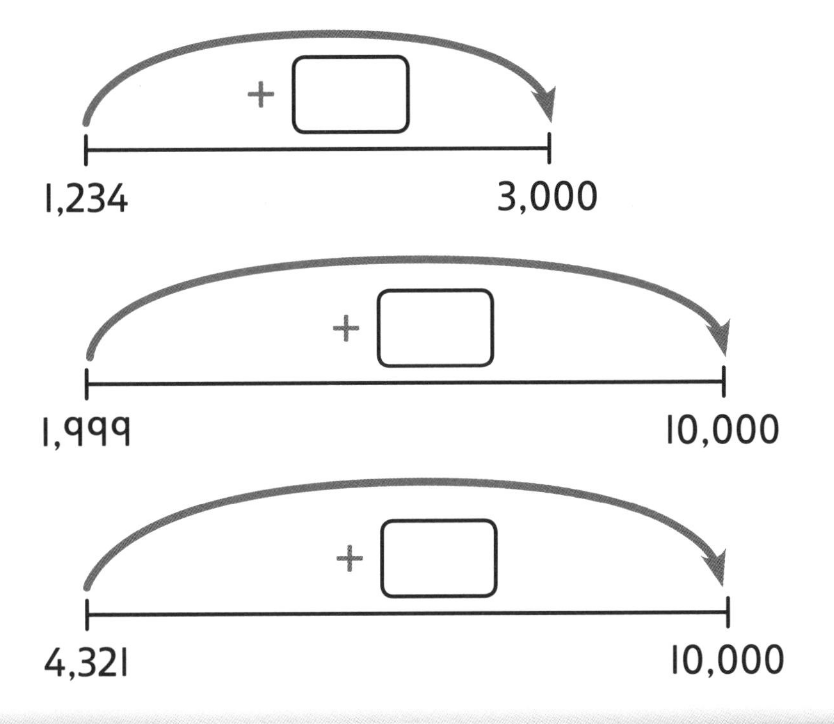

Reflect

When I add 4-digit numbers, I need to remember to:

1. ______________________________

2. ______________________________

3. ______________________________

→ Textbook 4A p104

Subtracting two 4-digit numbers 

1 A postal worker had 4,325 letters. She delivered 2,114 in the morning. How many did she have to deliver in the afternoon?

Th	H	T	O
1,000 1,000 1,000 1,000	100 100 100	10 10	1 1 1 1 1

	Th	H	T	O
	4	3	2	5
–				

4,325 – ☐ = ☐

She had to deliver ☐ letters in the afternoon.

2 Match each subtraction to the correct equipment and then solve each subtraction.

Th	H	T	O
1,000 1,000 1,000 1,000	100 100	10 10 10 10 10	1 1

	Th	H	T	O
	4	2	5	0
–	1	1	4	0

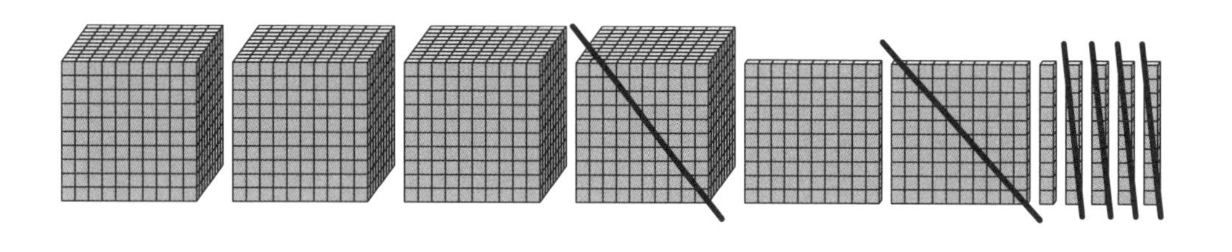

	Th	H	T	O
	4	5	2	5
–	2	1	1	4

Th	H	T	O

	Th	H	T	O
	4	2	5	2
–	2	0	1	1

3 Find the missing numbers.

a)

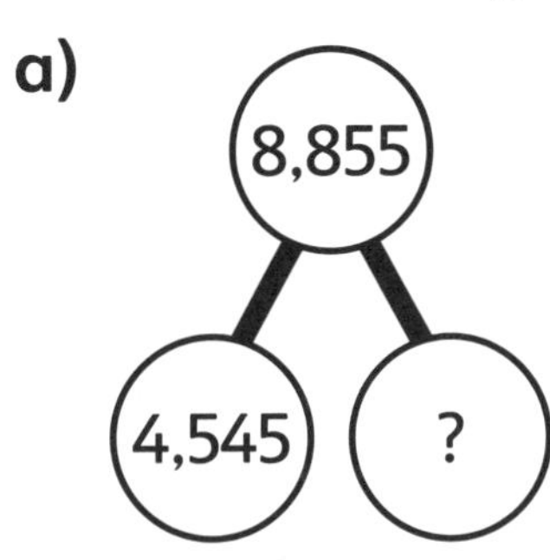

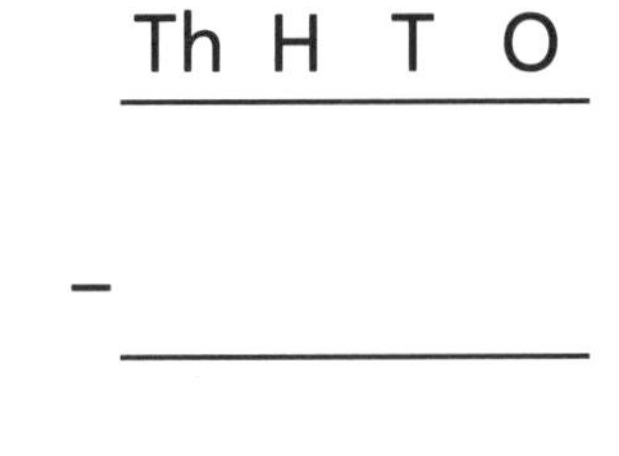

b)

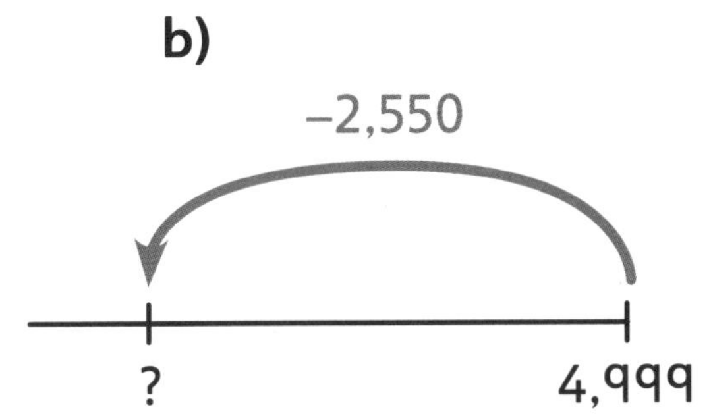

☐ – ☐ = ☐

	Th	H	T	O
–				

c)

9,099

2,066 ?

☐ – ☐ = ☐

	Th	H	T	O
–				

4 Explain the mistake.

9,732 – 411 = 5,622

	Th	H	T	O
	9	7	3	2
–	4	1	1	
	5	6	2	2

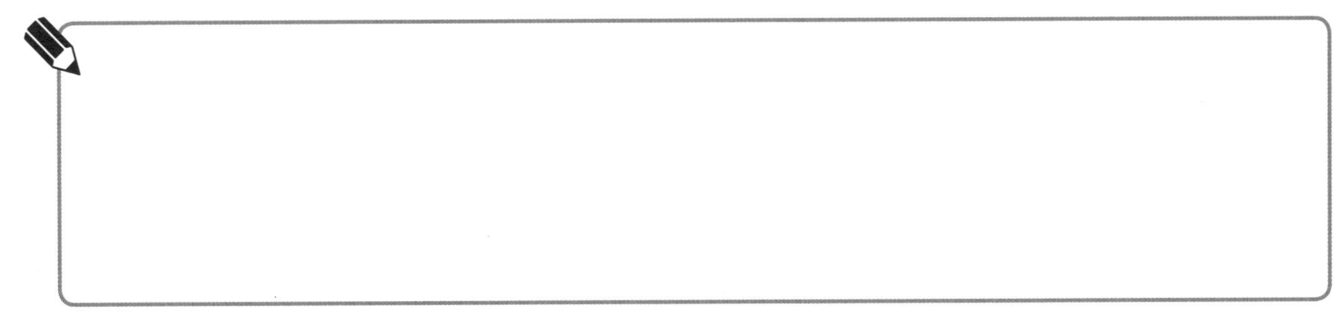

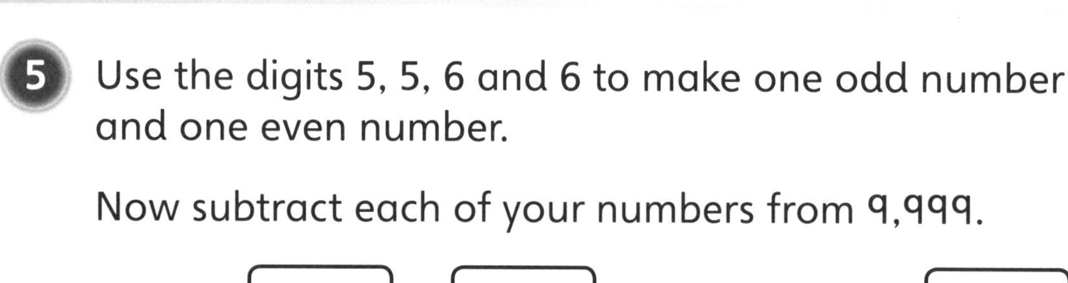

5 Use the digits 5, 5, 6 and 6 to make one odd number and one even number.

Now subtract each of your numbers from 9,999.

9,999 – ☐ = ☐ 9,999 – ☐ = ☐

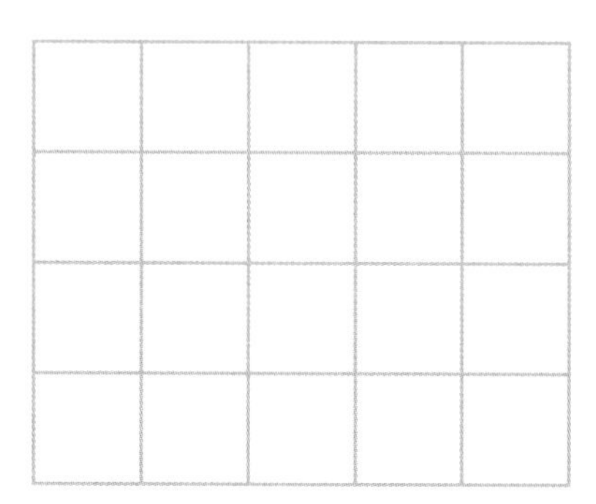

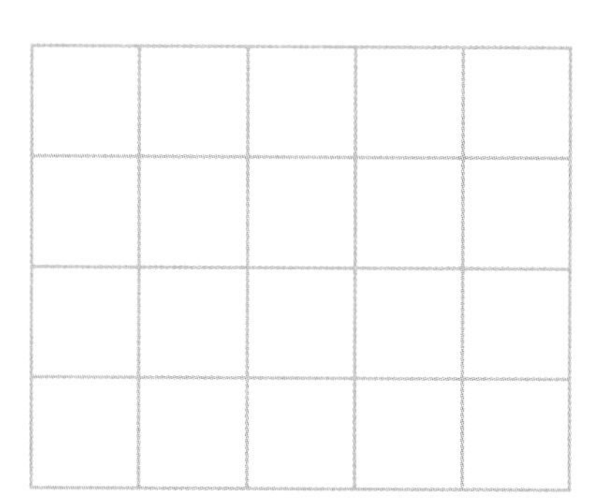

What do you notice about whether the answers are odd or even?

I noticed that ______________________________

____

____

____

Reflect

Write and solve a story problem for 5,455 – 2,123.

____

____

____

Subtracting two 4-digit numbers 

1 Complete the subtractions.

a) 4,362 – 247 = ☐

Th	H	T	O
1,000 1,000 1,000 1,000	100 100 100	10 10 10 10 10 10	1 1 1 1 1 1 1 1 1 1 1 1

Th	H	T	O
4	3	6	2
–	2	4	7

b) ☐ = 1,454 – 1,270

Th	H	T	O
1,000	100 100 100 100	10 10 10 10 10	1 1 1 1

Th	H	T	O
1	4	5	4
– 1	2	7	0

c) 2,350 – 1,530 = ☐

Th	H	T	O
1,000 1,000	100 100 100	10 10 10 10 10	

Th	H	T	O
–			

2 Jon wants to visit his friends. Kate lives 349 miles away. Bella lives 1,356 miles away. How much further away does Bella live than Kate?

Bella lives ☐ miles further away.

Th	H	T	O
–			

3 Complete these subtractions.

a) 9,375 – 8,293 = ☐

	Th	H	T	O
–				

c) 9,375 – 8,239 = ☐

	Th	H	T	O
–				

b) ☐ = 8,375 – 8,293

d) 7,375 – 239 = ☐

4 Find the missing numbers. 

a)

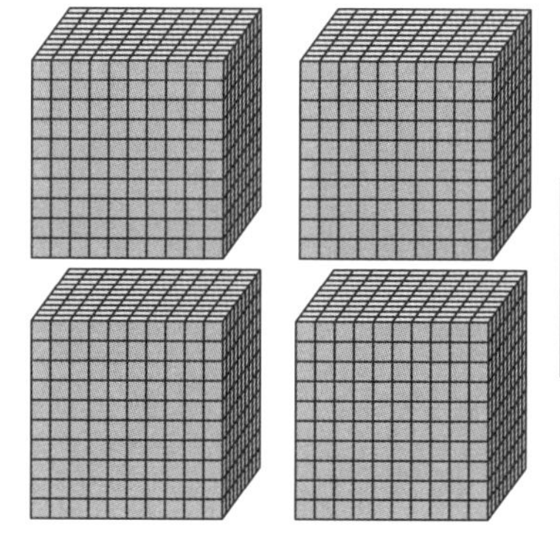

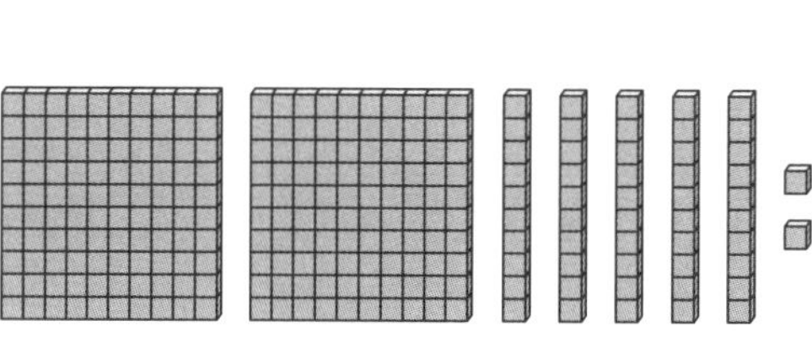

	Th	H	T	O
	4	2	5	2
–				
	2	1	1	3

b)

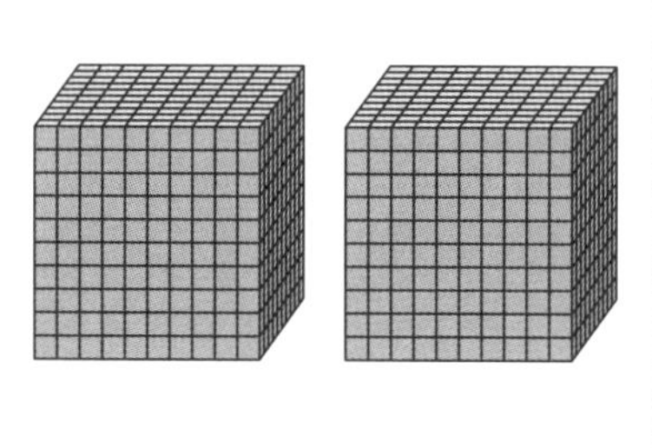

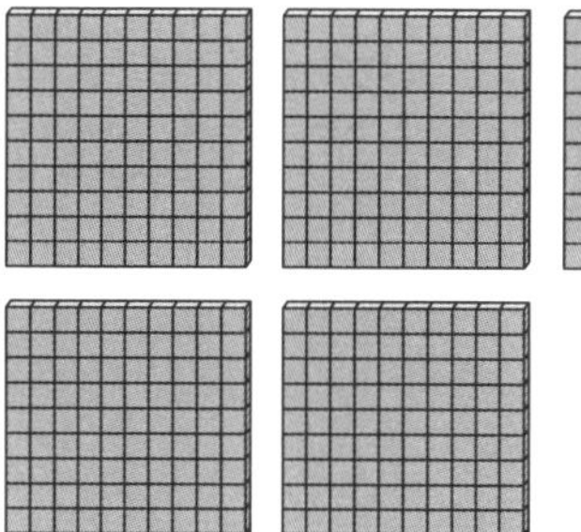

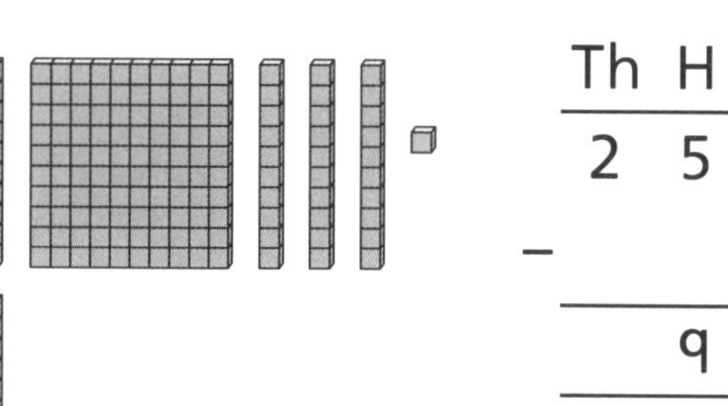

	Th	H	T	O
	2	5	3	1
–				
		9	1	1

5 Show the mental method you would use for each of these calculations.

a) 3,251 – 6 = ☐

3,251

c) 3,251 – 3,246 = ☐

3,251

b) 5,051 – ☐ = 4

d) 4,982 = 4,991 – ☐

Reflect

Write and solve a subtraction that needs an exchange of 1 hundred for 10 tens.

→ Textbook 4A p112

Subtracting two 4-digit numbers 3

1 Max scored 2,335 points in a game. Isla scored 418 fewer points. How many points did Isla get?

Th	H	T	O
1,000 1,000	100 100 100	10 10 10	1 1 1 1 1

	Th	H	T	O
	2	3	3	5
–		4	1	8

2,335 – 418 = ☐

Isla got ☐ points.

2 Complete these subtractions.

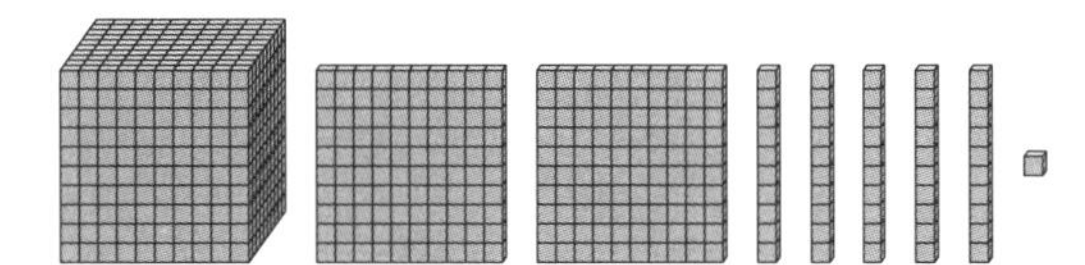

a)

	Th	H	T	O
	1	2	5	1
–		1	8	2

1,251 – 182 = ☐

c)

	Th	H	T	O
	2	2	9	2
–	1	1	9	9

2,292 – 1,199 = ☐

b)

	Th	H	T	O
–				

3,150 – 225 = ☐

d)

	Th	H	T	O
–				

☐ = 3,150 – 1,160

3 **a)** Solve 9,449 – 777.

Th	H	T	O
–			

9,449 – 777 = ☐

b) Explain how you can use this to work out 8,449 – 777.

8,449 – 777 = ☐ because...

____

____

____

____

4 Complete the table.

Subtraction	**Number of exchanges**	**Solution**
1,258 – 163	1	Th H T O 1 2 5 8 – 1 6 3 ______
3,258 – 329	☐	Th H T O – ______
1,☐58 – 24☐	2	Th H T O – ______

5 Explain the mistake and show the correct calculation.

	Th	H	T	O
	3	4	1	2
−	1	6	5	1
	2	2	4	1

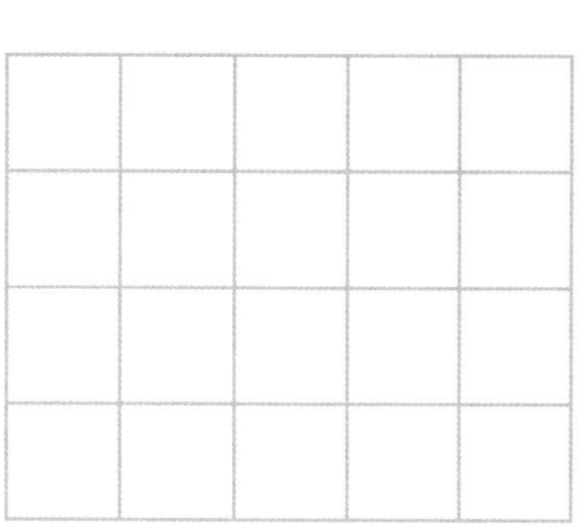

The mistake is that ______________________________

____

____

6 Richard thinks that the rabbit's weight is closer to the cat's weight than it is to the guinea pig's weight.

Do you agree with Richard? Explain your answer.

____

____

Reflect

Explain how you can tell if you need zero, one or two exchanges in a subtraction.

____

____

→ Textbook 4A p116

Subtracting two 4-digit numbers 

1 Olivia is reading a story that is 1,401 words long. She has read 225 words so far. How many words does she have left to read?

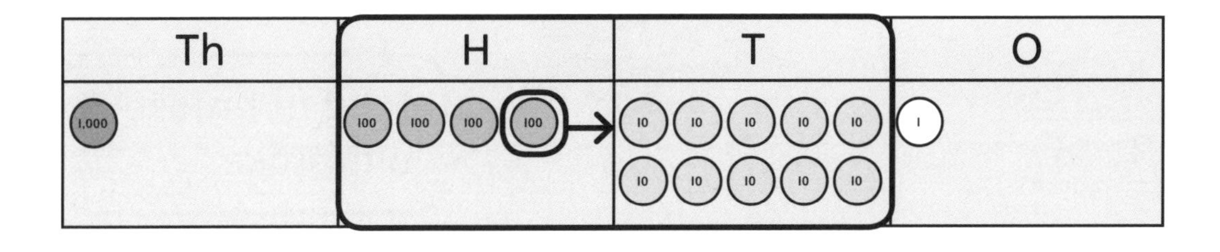

Th	H	T	O
1,000	100 100 100	10 10 10 10 10 10 10 10 10 10	1 1 1 1 1 1 1 1 1 1 1

	Th	H	T	O
	1	4	0	1
−		2	2	5

[ ] − [ ] = [ ]

She still has [ ] words left to read.

2 Draw place value counters to show the exchanges that need to be made. Complete the subtraction.

Th	H	T	O

	Th	H	T	O
	2	2	0	2
−		1	5	7

a) Join each subtraction to the statement that describes it.

3,507 – 419 = 3,198

	Th	H	T	O
	3	5	$^{1}0$	$^{1}7$
–		4	1	9
	3	1	9	8

3,008 – 1,419 = 1,599

	Th	H	T	O
	$^{2}\not{3}$	$^{9}\,{}^{1}\not{0}$	$^{1}0$	$^{1}8$
–	1	4	1	9
	1	5	9	9

3,023 – 419 = 2,604

	Th	H	T	O
	$^{2}\not{3}$	$^{1}0$	$^{1}\not{2}$	$^{1}3$
–		4	1	9
	2	6	0	4

I did not make a mistake.

I just added an extra 10 ones and 10 tens, but I did not exchange.

I did not correctly exchange 1 ten for 10 ones.

b) Correct the two calculations that have mistakes.

☐ – ☐ = ☐ ☐ – ☐ = ☐

	Th	H	T	O
–				

	Th	H	T	O
–				

4 **a)** Jake has partitioned a number to do a subtraction. Write the subtraction to match it.

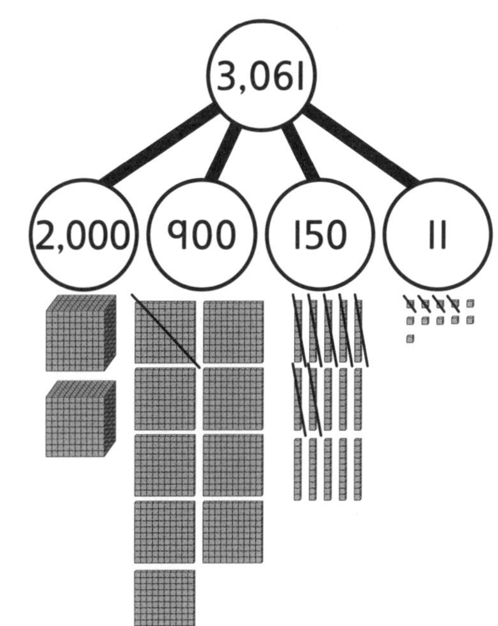

	Th	H	T	O
−				

☐ − ☐ = ☐

b) Complete the partition to match this subtraction.

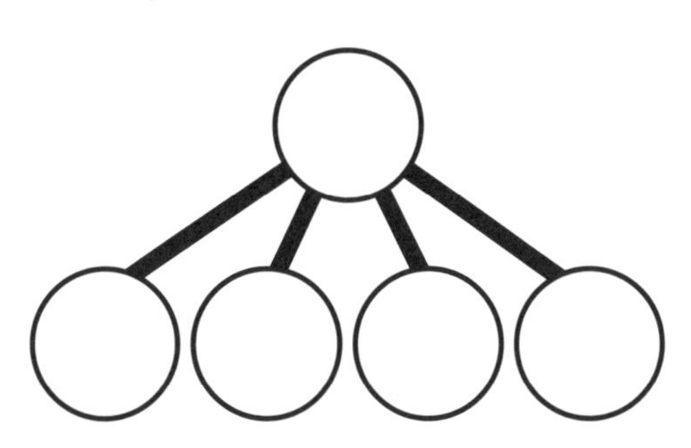

	Th	H	T	O
	3	5	0	1
−	2	5	5	2

☐ − ☐ = ☐

Reflect

If I need to exchange 10 ones when there is a zero in the tens column, I ...

→ Textbook 4A p120

Equivalent difference

1 Write a subtraction to go with each model. Complete all the subtractions. Circle the one you chose to solve first.

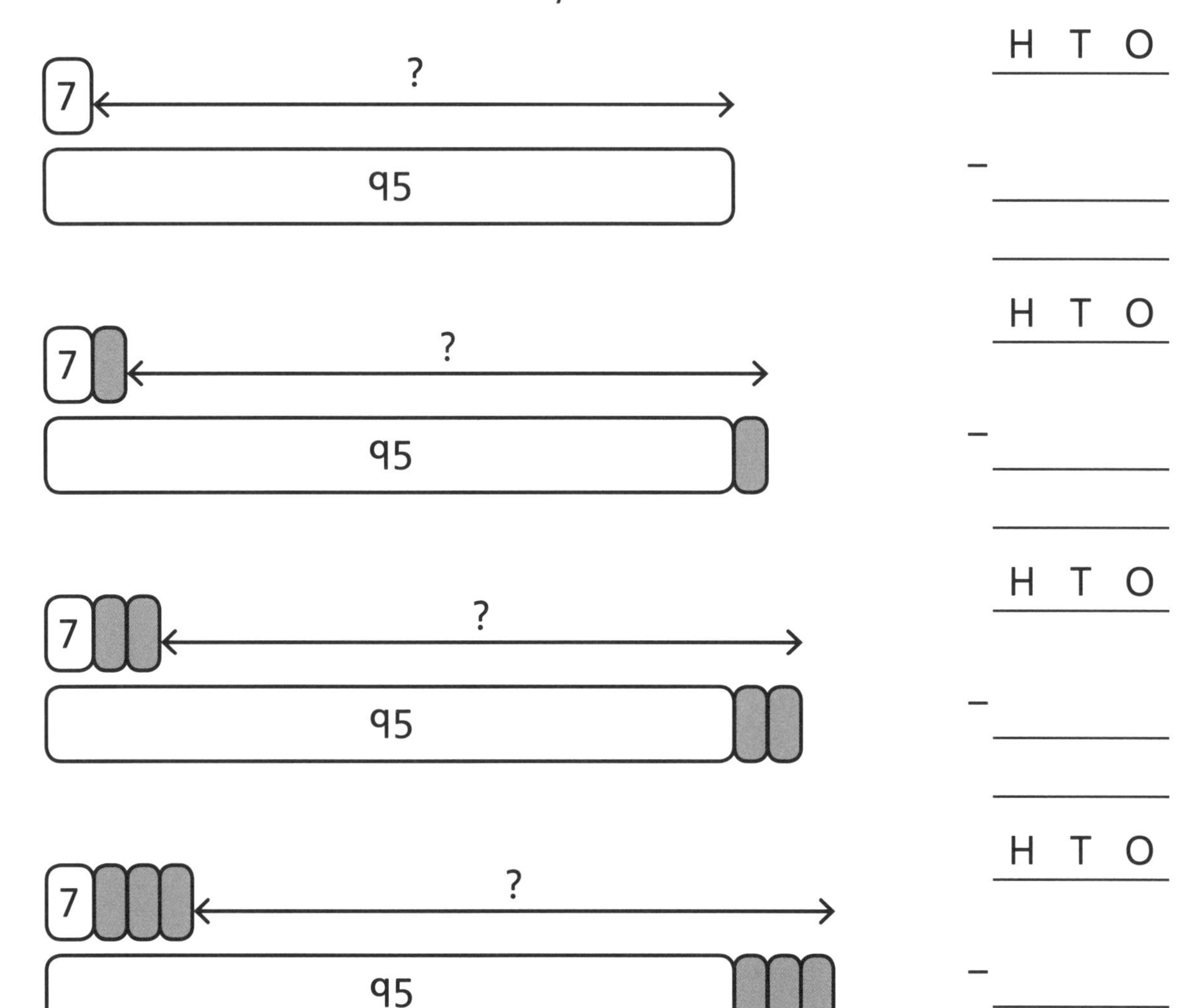

2 Write a subtraction to solve 298 – 139 = ☐.

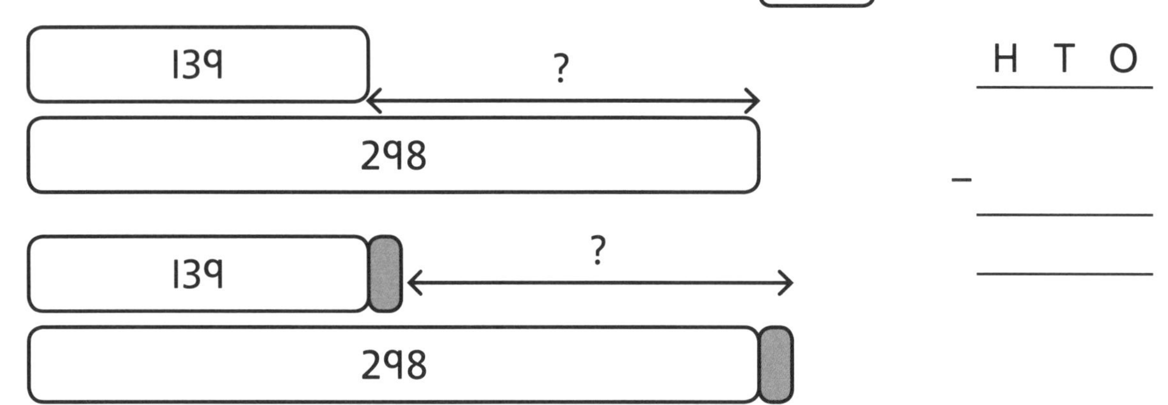

3 Jan's tower is 235 cm tall. Anne's is 98 cm tall. Write subtractions to find the difference between the height of the towers. Circle the one you choose to complete first.

	H	T	O
	2	3	5
–		9	8

	H	T	O
	2	3	6
–			

	H	T	O
–			

	H	T	O
–			

	H	T	O
–			

☐'s tower is ☐ cm taller.

4 **a)** Ebo solved 2,001 – 567 = ☐ with the calculation 1999 – 565 = ☐.

Complete his calculation to find the answer.

	Th	H	T	O
	1	9	9	9
–		5	6	5

b) Choose one of these subtractions to solve with a similar method.

1,507 – 385 = ☐ 1,000 – 518 = ☐

I chose ☐ – ☐ because ______________________________

____

Now solve the subtraction. Show your method.

5 Choose a method to use to solve each of these subtractions. Think about which method is the most efficient each time.

2,950 – 850

2,875 – 1,989

3,011 – 2,997

8,001 – 4,567

6,626 – 6,618

9,009 – 10

Reflect

Think of another method to solve 1,000 – 955. Discuss with your partner which you think is most efficient.

	Th	H	T	O
	1	0	0	0
–		9	5	5

I think the best method is to ______________________________

____

____

because ____________________________________

____

____

→ Textbook 4A p124

Estimating answers to additions and subtractions

1 Round to the nearest 1,000 to estimate these calculations.

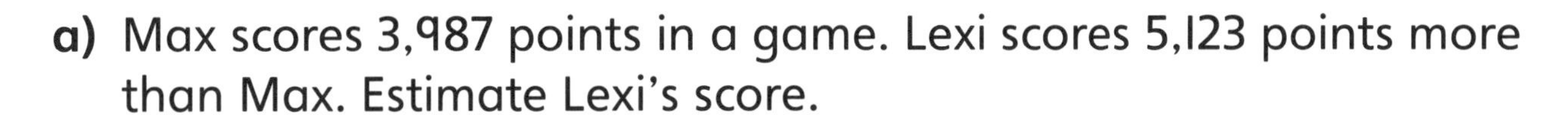

a) Max scores 3,987 points in a game. Lexi scores 5,123 points more than Max. Estimate Lexi's score.

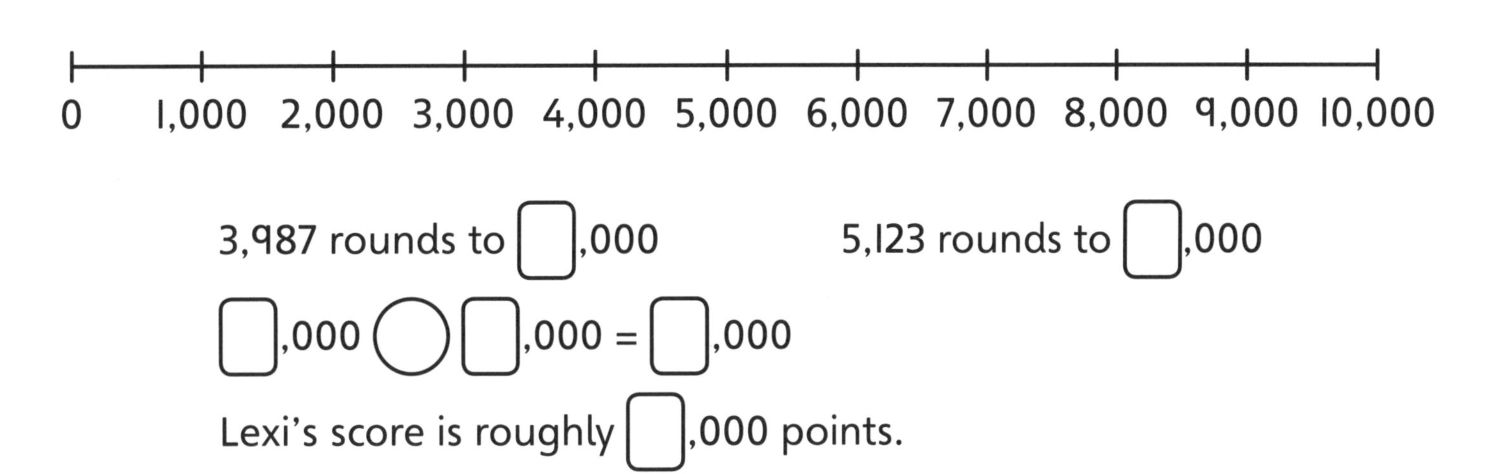

3,987 rounds to ☐,000 5,123 rounds to ☐,000

☐,000 ◯ ☐,000 = ☐,000

Lexi's score is roughly ☐,000 points.

b) Max loses 3,104 points. Estimate how many points he has now.

☐,000 – ☐,000 = ☐,000 Max has roughly ☐ points now.

c) Now work out the exact scores and compare them with your estimates.

Lexi's exact score

Th	H	T	O

Max's exact score

Th	H	T	O

Are your estimates close to the exact answers?

____

2 Join each calculation to the estimate that best matches it.

Some of the estimates do not have a good match, and some of the estimates match to more than one calculation.

Calculations
2,101 – 998
2,891 – 1,100
1,975 + 2,010
1,998 + 2,101
2,925 – 975
2,998 – 1,998

Estimates
2,000 + 2,000
2,900 – 1,000
3,000 – 2,000
2,100 – 1,000
1,000 + 2,000
3,000 + 2,000

3 **a)** Complete each calculation. Then write an estimate to check.

6,152 + 3,025 = ☐

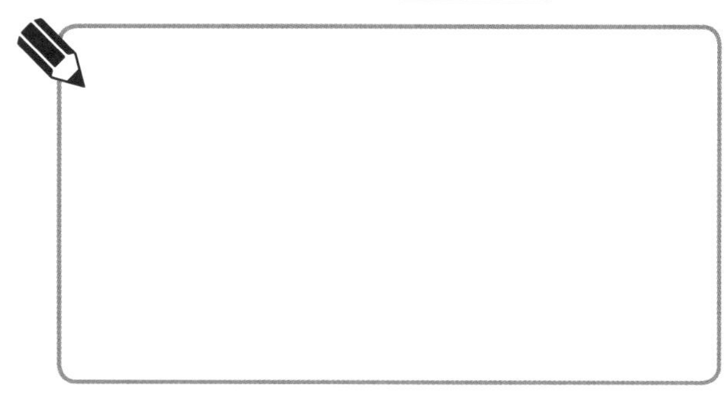

Estimate:

☐ + ☐ = ☐

6,452 – 2,005 = ☐

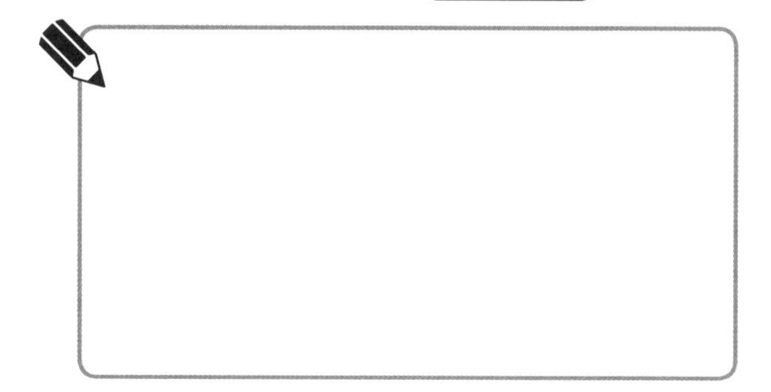

Estimate:

b) Explain why you chose each of your estimation methods.

____

____

____

4 6,491 – 2,725 = ☐

Try rounding the numbers to the nearest 1,000 to estimate the answer. Then estimate by rounding to the nearest 100. Then estimate by rounding to the nearest 10.

Nearest 1,000	Nearest 100	Nearest 10
Estimate: ☐	Estimate: ☐	Estimate: ☐

Find the exact answer and compare it to each of your estimates.

What do you notice?

Th H T O

____

____

____

Reflect

Explain how you would estimate 1,915 – 1,019.

____

____

____

→ Textbook 4A p128

Checking strategies

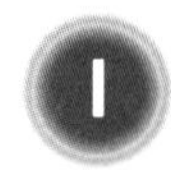

a) Check Emma's subtractions using the inverse operation, and give each a tick in the box if it is correct or a cross if it is wrong.

3,412 – 1,151 = 2,341

1,001 – 550 = 451

9,876 – 6,789 = 2,189

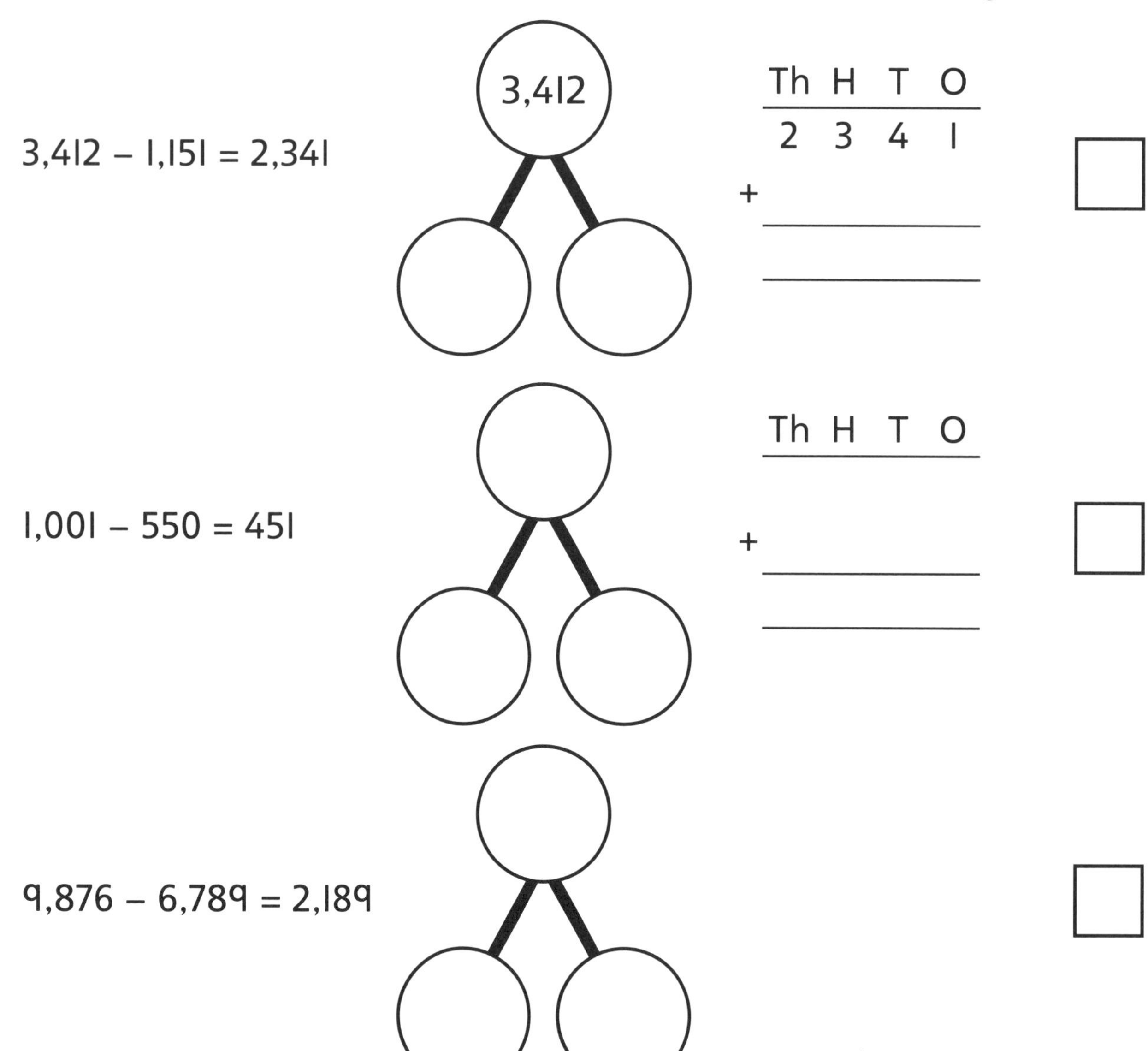

b) Write the correct subtractions.

2 Holly bought a car for £1,899. She also paid £995 to get it repaired. Holly has calculated that she spent £2,894 in total.

Check Holly's calculation.

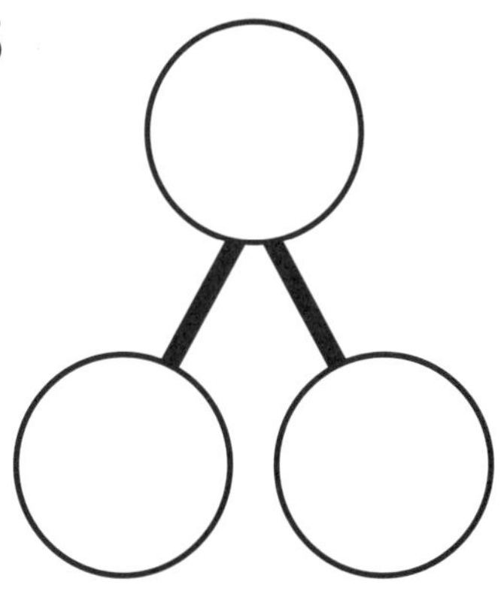

I think Holly is correct / incorrect because ____________________

____

____

____

3 Calculate the missing numbers.

a) ☐ + 995 = 5,555

b) ☐ – 5,555 = 995

c) 5,555 – ☐ = 995

d) ☐ – 995 = ☐

4,499 + 3,499 = 7,998 

Do you agree with Dexter that his estimate is not right?

Explain how you would check this calculation.

I rounded to estimate 4,000 + 3,000 = 7,000, but the answer rounds to 8,000 so I do not think my estimate is right.

5 Find the answer to this calculation and then show one way to check your answer.

CHALLENGE

Isla takes 2,599 paper clips out of a bag. There are 2,599 paper clips left in the bag. How many paper clips are there altogether?

Reflect

Show how to check 599 + 1,599 = 2,098 using both estimating and an inverse operation.

Problem solving – addition and subtraction 1

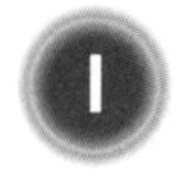

a) Ambika poured 2,500 ml of water onto a flower bed. Aki poured 3,100 ml of water. How much water did they pour altogether?

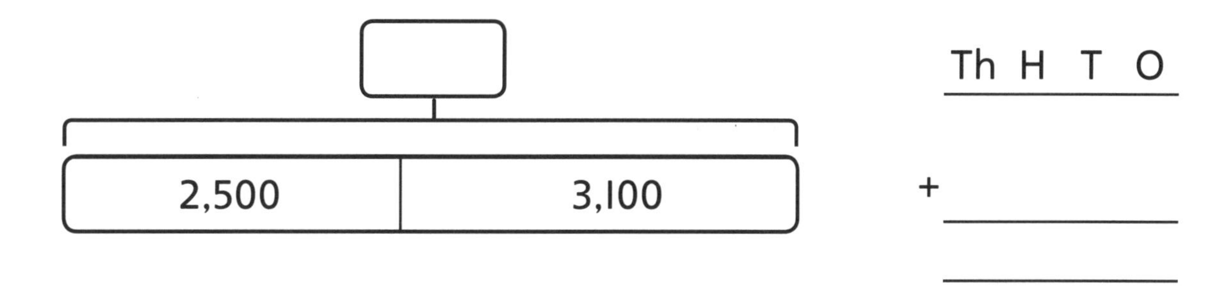

They poured ☐ ml of water altogether.

b) Ambika started with 5,000 ml in her watering can. How much water does she have left now?

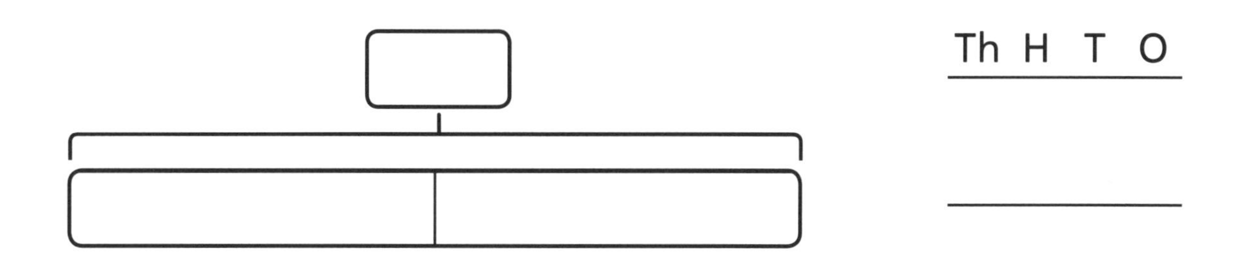

Ambika has ☐ ml of water left now.

2 Complete bar models to show both of these problems, then find the solutions to them.

a) Mrs Dean lives 5,000 m from her school. She has cycled 3,900 m so far. How far does she have left to cycle?

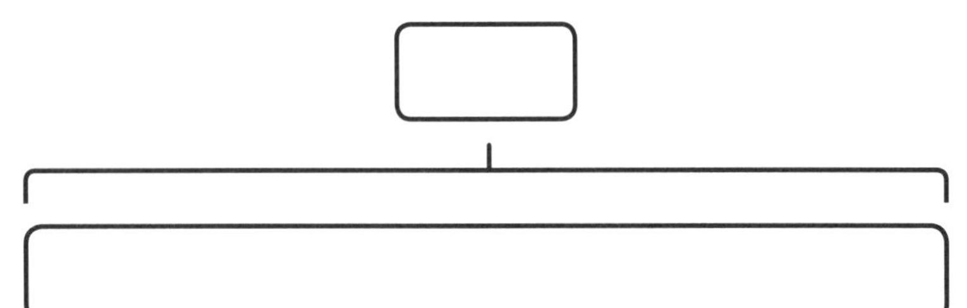

Th	H	T	O

She has ☐ m left to cycle.

b) Mr Jones walks 1,250 m to the bus stop, then travels 2,800 m on the bus. How far does he travel altogether?

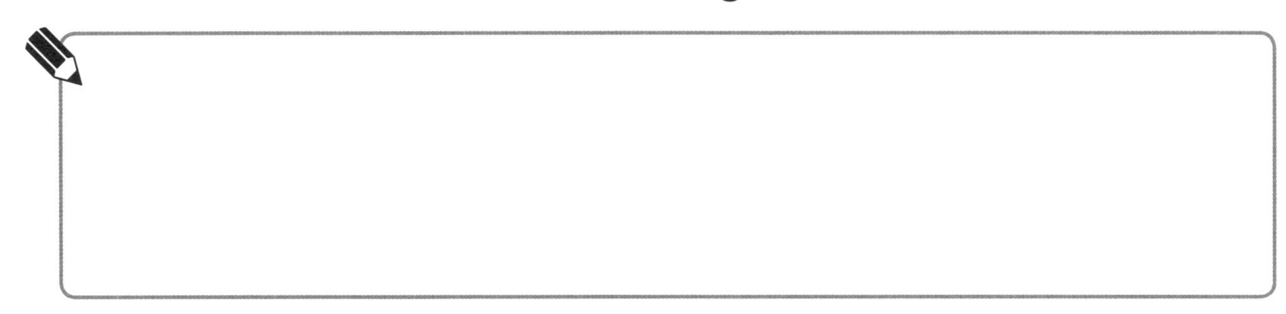

He travels ☐ m altogether.

3 Draw bar models and find the missing numbers.

a) ☐ – 3,750 = 4,000 **b)** 4,000 – ☐ = 3,750

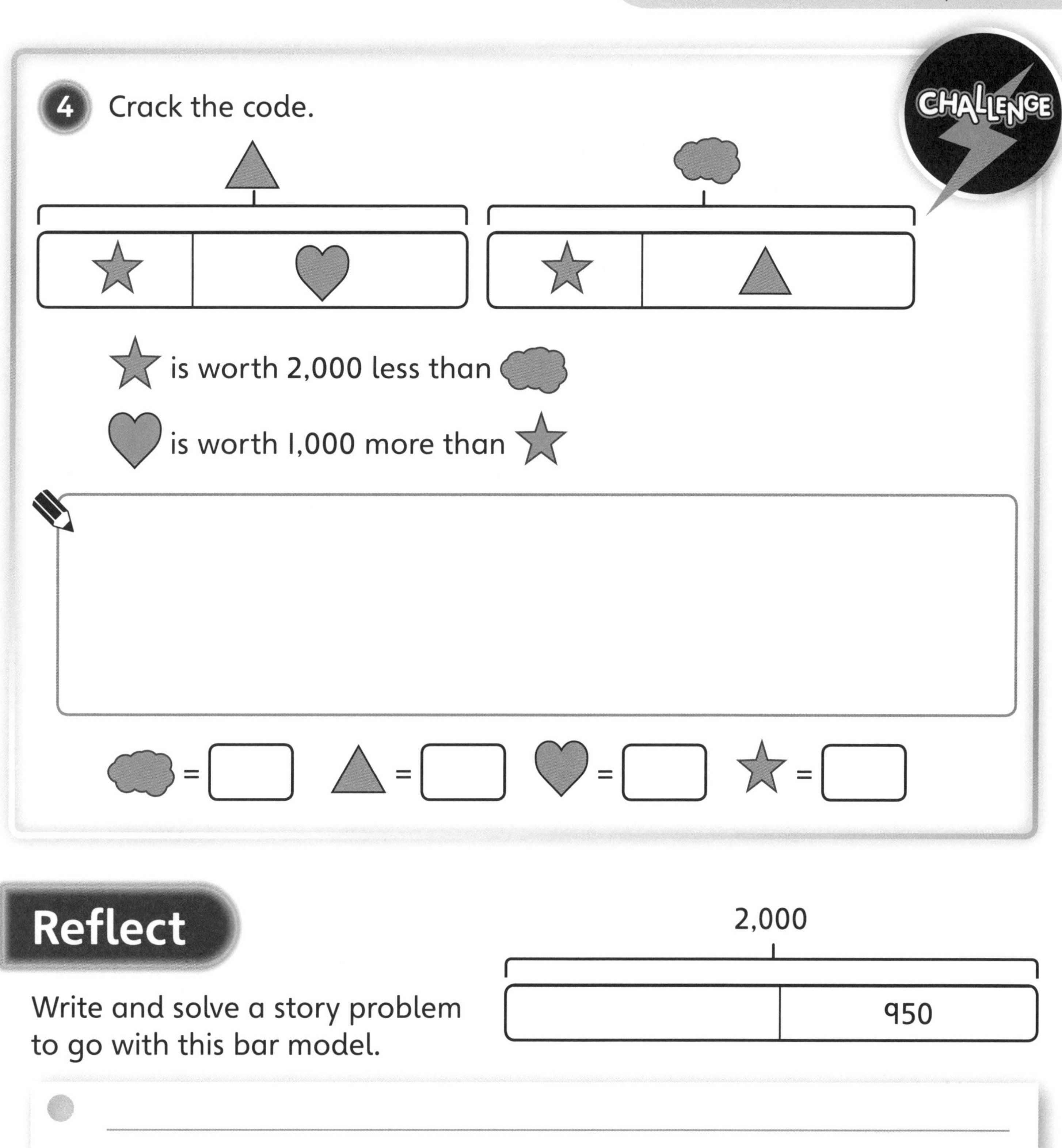

Reflect

Write and solve a story problem to go with this bar model.

2,000	
	950

→ Textbook 4A p136

Problem solving – addition and subtraction 2

1 a) Ebo has 1,020 football stickers. Richard has 820 football stickers.

How many more stickers does Ebo have?

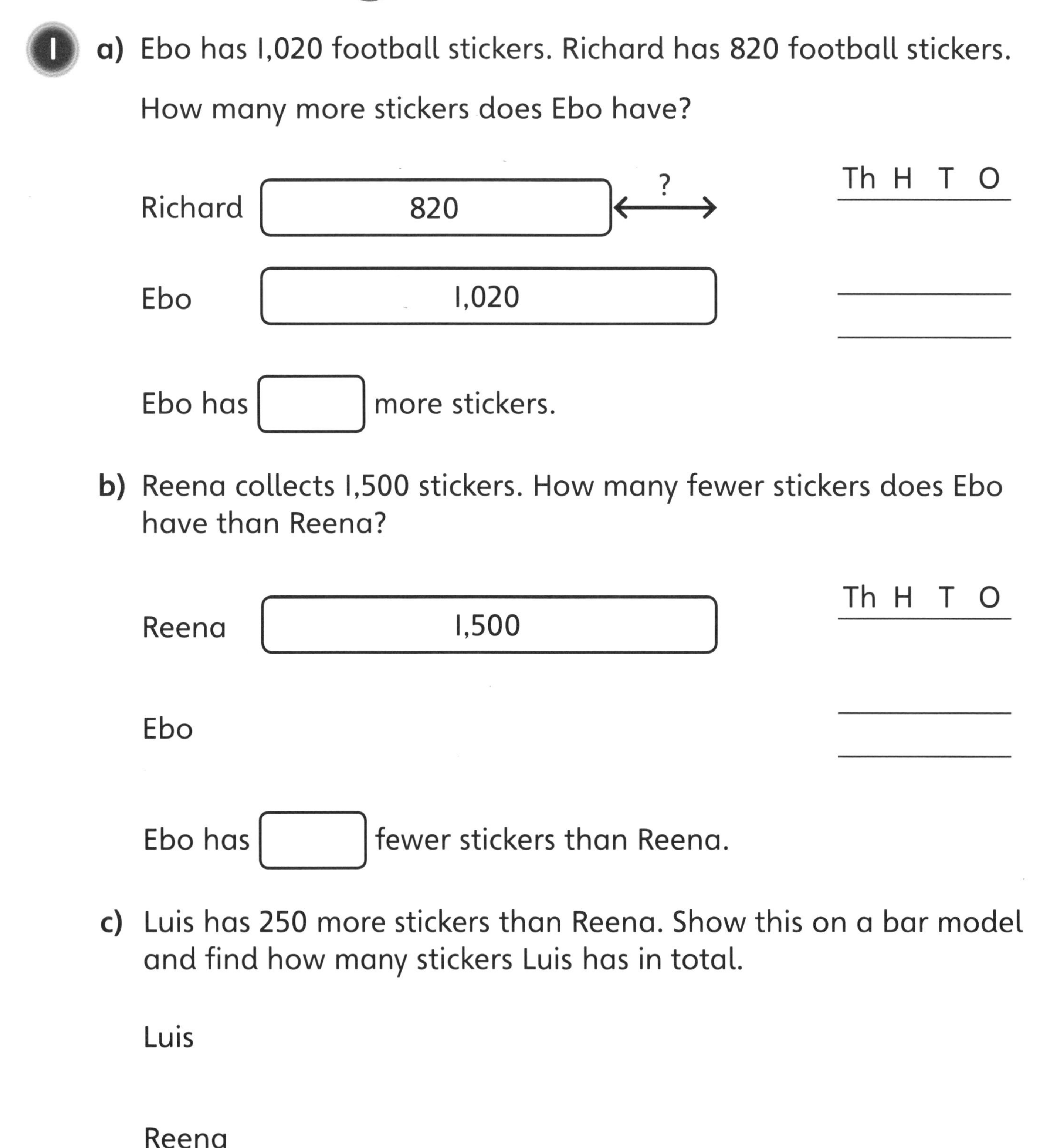

Ebo has ☐ more stickers.

b) Reena collects 1,500 stickers. How many fewer stickers does Ebo have than Reena?

Ebo has ☐ fewer stickers than Reena.

c) Luis has 250 more stickers than Reena. Show this on a bar model and find how many stickers Luis has in total.

Luis has ☐ stickers in total.

2 Mo collects 425 shells and Lee collects 576 shells. How many more shells does Lee collect?

Explain which bar model suits this problem.

A
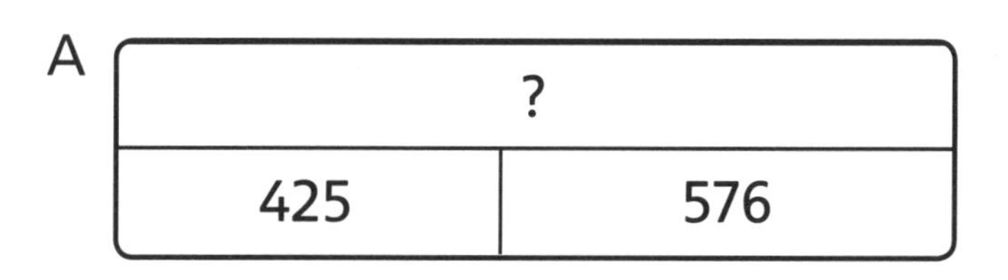

B
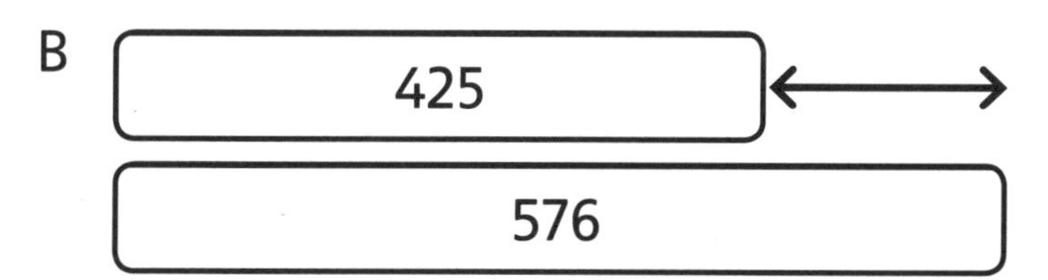

I think A / B suits this problem best because ______________________________

____

____

____

3 Draw a bar model and solve this problem.

Max has 1,500 ml of paint. Isla has 750 ml more paint than Max.

Max uses 500 ml of paint. Isla also uses some paint and now they have the same amount of paint left as each other.

How much paint did Isla use?

CHALLENGE

4 Solve this story problem by drawing bar models.

Bella, Aki and Andy each think of a number.

Bella's number is 875 more than Aki's number.

Aki subtracts 499 from his number. Now Aki's number is 245 less than Andy's number.

What is the difference between Bella's number and Andy's number?

The difference between Bella's number and Andy's number is ☐.

Reflect

I would draw a comparison bar model when ______________________

____

I would draw a single bar model when ______________________

____

→ Textbook 4A p140

Problem solving – addition and subtraction 3

a) Sofia entered a triathlon. She swam 500 m, cycled 2,250 m and ran 1,250 m to the finish. What was the total distance?

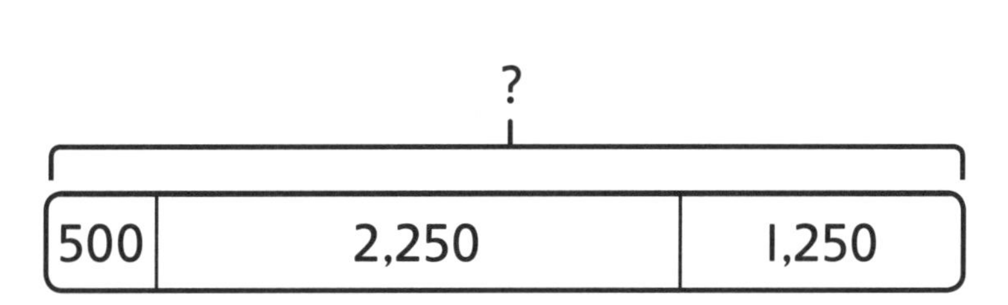

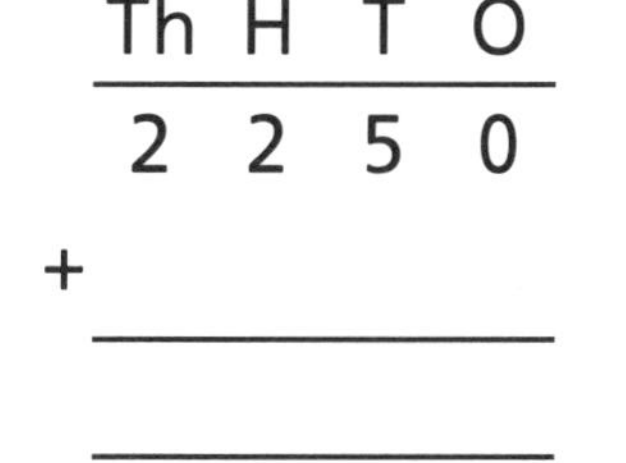

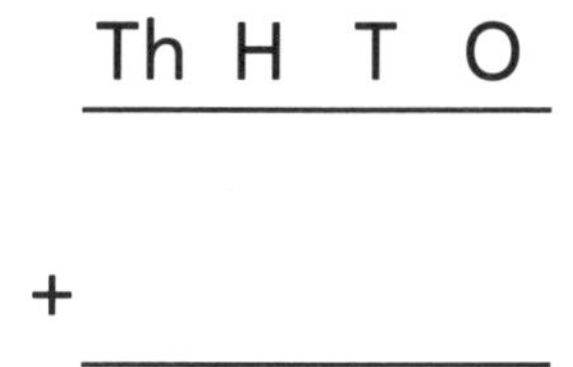

The total distance was ☐ m.

b) Mrs Dean entered an 8,000 m triathlon. She ran 2,500 m and cycled 4,750 m. How far did she swim?

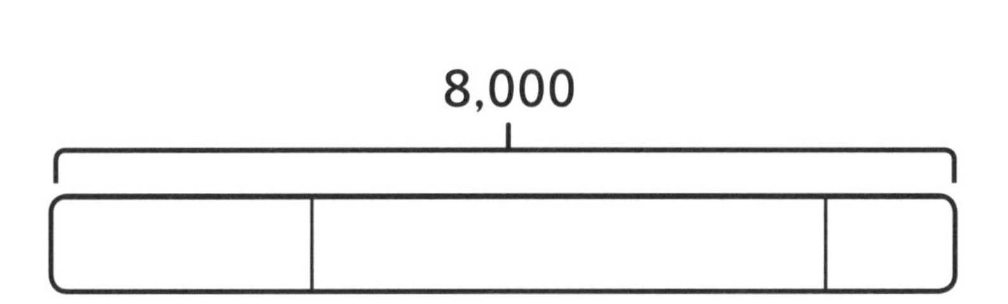

Th H T O

Th H T O

c) Explain the order you chose to do the calculations in for part **b)**.

2 What is the height of the middle section of the tower?

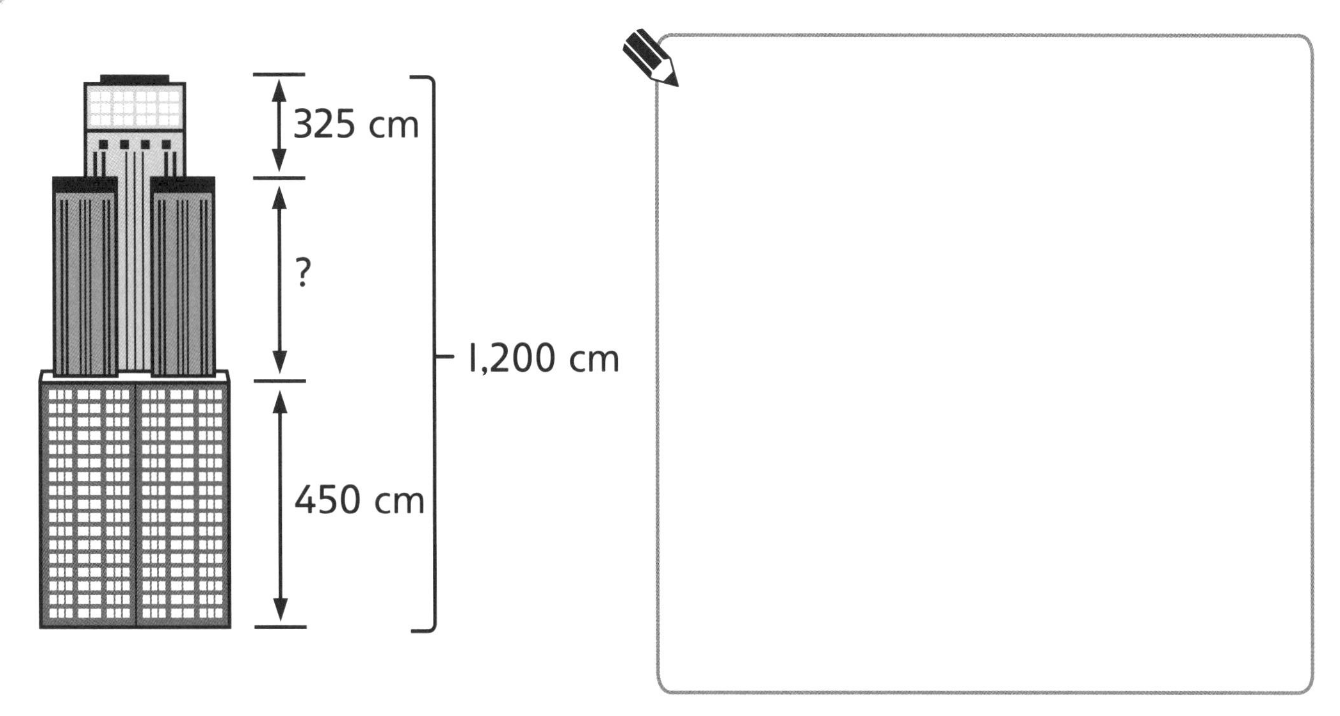

The height of the middle section of the tower is [ ] cm.

3 Draw a bar model and solve this story problem.

There are 650 children in a primary school. There are 1,100 more children in the secondary school. How many children are there in total?

CHALLENGE

4 a) Amy has £1,275 less than her brother Ben. Then Ben spends £550 and Amy gets £750. Who has more money now? What is the difference between the amounts that Amy and Ben now have?

_______________ has more money now.

The difference is £ [ ] .

b) Evelyn has £800 more than Noah. Together they have £2,800. How much do they each have?

Evelyn has £ [ ] and Noah has £ [ ] .

Reflect

Draw a bar model with three parts that total 2,050.

→ Textbook 4A p144

Problem solving – addition and subtraction 4

1 Mr Jones's school collected 5,000 bottles for a recycling competition.

- Class 1 collected 1,228 bottles.
- Class 3 collected 1,517 bottles.
- Class 4 collected 483 bottles.
- Class 2 think they collected the most bottles.

a) Complete both diagrams to show this problem.

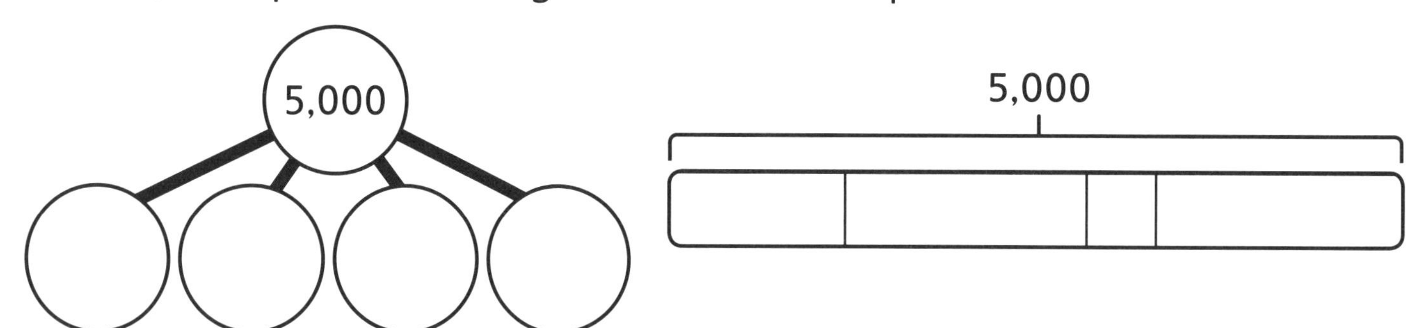

b) Calculate how many bottles Class 2 collected. Which class collected the most bottles?

Th H T O   Th H T O   Th H T O

Class 2 collected ☐ bottles.

☐ bottles < ☐ bottles < ☐ bottles < ☐ bottles

Class ☐ collected the most bottles.

2 There are 3,985 United fans at a football match and 1,700 fewer Rovers fans. How many fans are there in total?

United [3,985]

Rovers [ ] ← [ ] →

?

There are [ ] fans in total.

3 A rabbit weighs 1,502 g. A hamster weighs 4,586 g less than a small dog. The dog weighs 3,116 g more than the rabbit. How much does the hamster weigh?

The hamster weighs [ ] g.

CHALLENGE

4 Write a story problem to match the diagram.

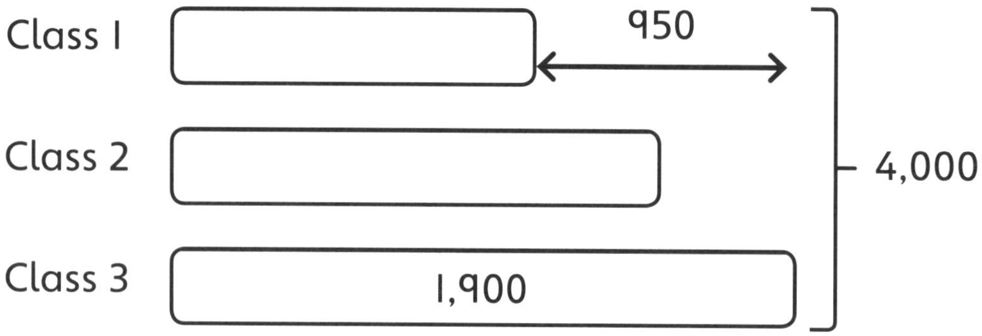

Class 1 | 950
Class 2 | 4,000
Class 3 | 1,900

Reflect

When I draw a bar model to help me solve a problem, I decide how many bars I need to draw by

→ Textbook 4A p148

End of unit check

1. 1,849 + [ ] = 8,634   2,026 = 9,000 − [ ]

Isla knows that one of these calculations has a missing number greater than 6,800, but she cannot remember which one it is.

Make a prediction and explain how you chose it.

Then show how to complete each calculation accurately.

2 Aki, Jamilla and Lee are playing a game.

Aki scores 4,875 points.

Lee scores 8,699 points.

Jamilla scores 3,823 less than Aki.

Aki thinks his score is closer to Lee's score than it is to Jamilla's score.

Explain whether or not Aki is correct. You may use diagrams to explain.

Power check

How do you feel about your work in this unit?

Power puzzle

What is the value of each shape?

Puzzle A

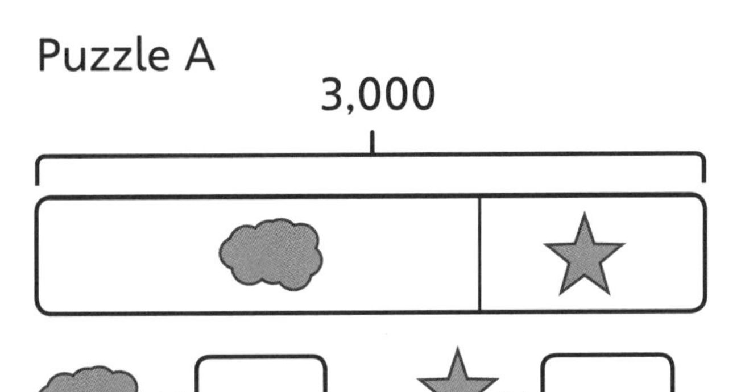

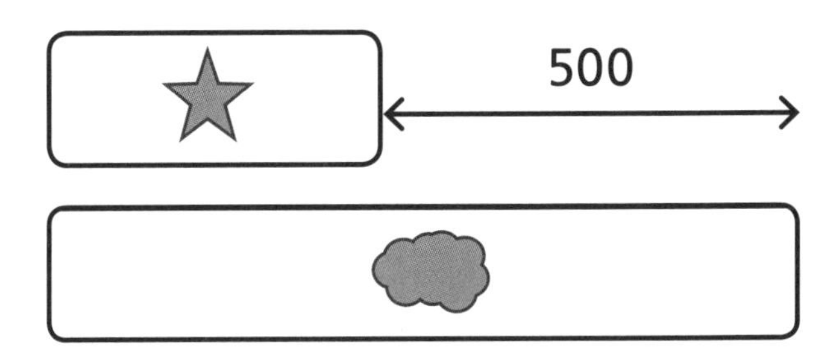

Puzzle B

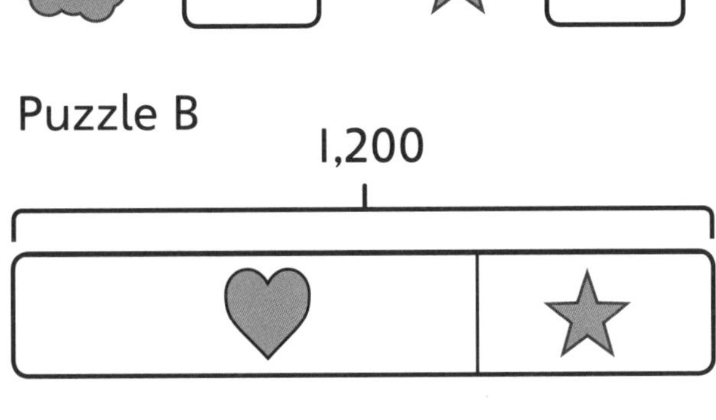

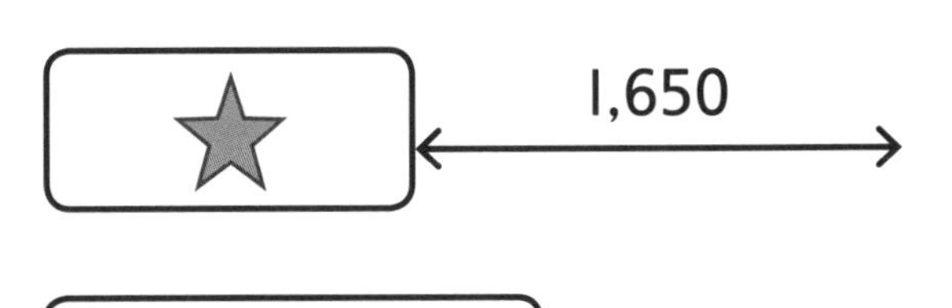

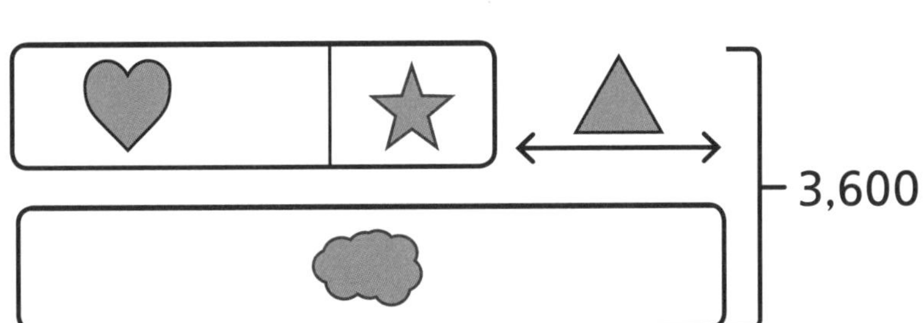

Create your own puzzle like this for your partner to solve. Choose two numbers and draw the bar model. Show your partner the total and the difference but hide the numbers from them.

→ Textbook 4A p152

Kilometres

Barwich 3 km

Littleton 6 km

Newville 9,000 m

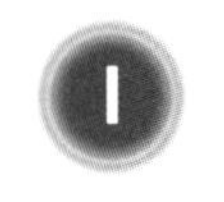

Complete the bar models to help you convert each distance.

a) How far away is Barwich in metres?

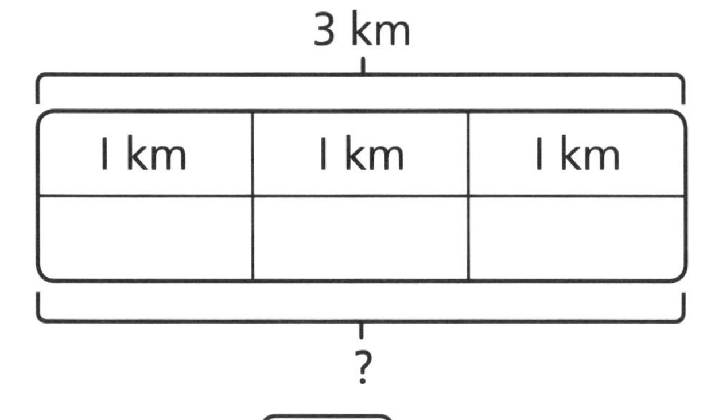

3 km = ☐ m

Barwich is ☐ metres away.

b) How far away is Littleton in metres?

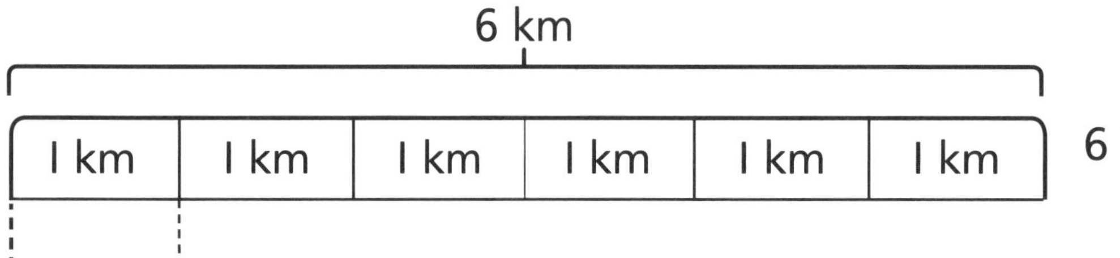

6 km = ☐ m

Littleton is ☐ metres away.

c) How far away is Newville in kilometres?

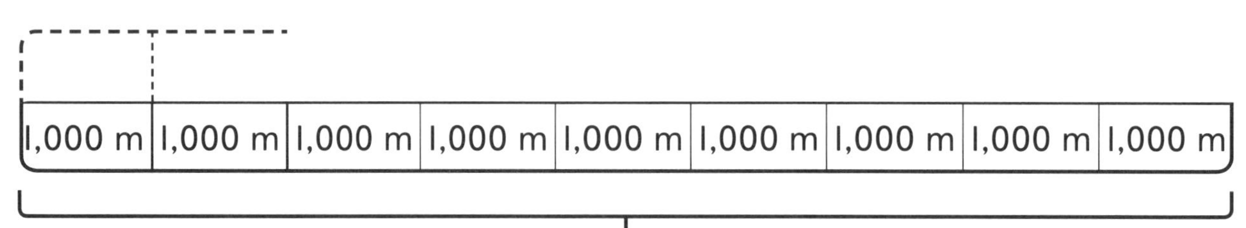

☐ km = 9,000 m

Newville is ☐ kilometres away.

2 Use these number lines to work out the equivalent distances.

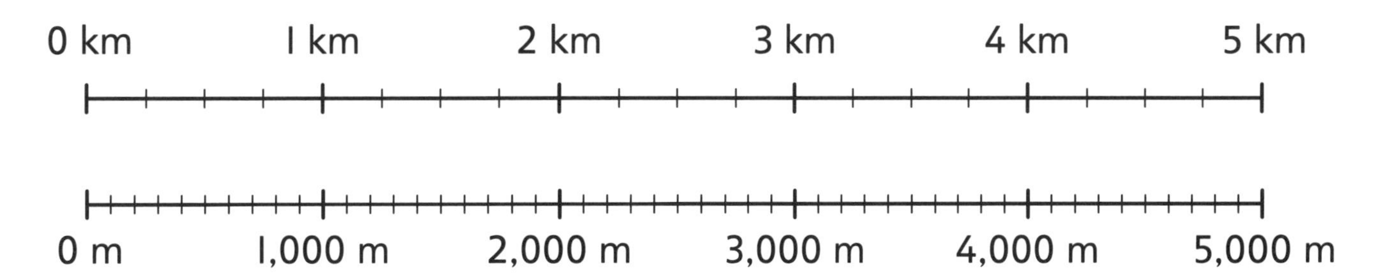

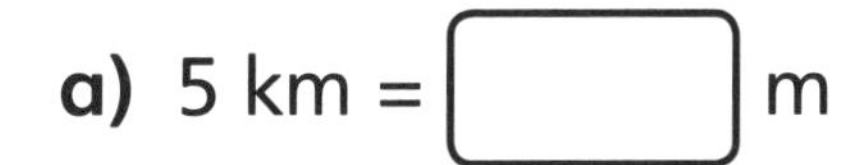

a) 5 km = ☐ m

c) ☐ m = $3\frac{1}{2}$ km

b) 1,500 m = ☐ $\frac{☐}{☐}$ km

d) ☐ $\frac{☐}{☐}$ km = 1,250 m

3 Complete the equivalent distances.

a) 11 km = ☐ m

c) ☐ km = 8,000 m

b) $4\frac{1}{2}$ km = ☐ m

d) ☐ km = 10,500 m

4 The town council is planting flowers beside $9\frac{1}{2}$ km of roads.

This costs £1 per metre. How much will it cost altogether?

Show your working in the box below.

The flowers will cost £ ☐.

5 The distance between any two stations is 1,000 m.

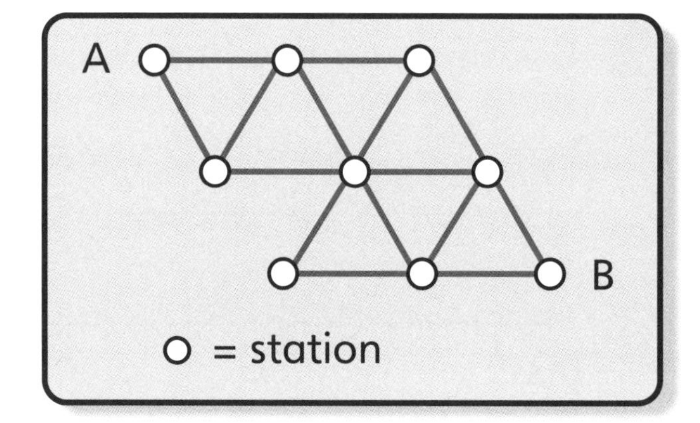

Draw a route from A to B on the map.

Write down the number of kilometres the train travels on the route.

☐ km

6 Write these distances in metres.

CHALLENGE

a) $\frac{1}{2}$ km = ☐ m

b) $\frac{3}{4}$ km = ☐ m

c) $\frac{2}{5}$ km = ☐ m

d) $\frac{1}{4}$ km = ☐ m

e) $\frac{1}{5}$ km = ☐ m

f) $\frac{1}{10}$ km = ☐ m

I can use what I know about kilometres to work out fractions of kilometres in metres.

Reflect

2,000 m + 500 m + 1 km = ☐ km

Use what you have learnt to explain how you would work out the answer.

→ Textbook 4A p156

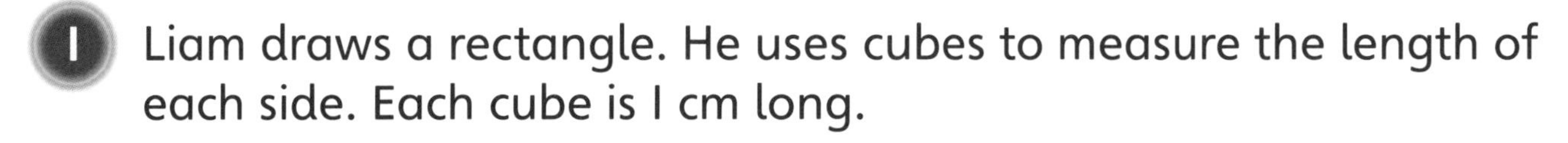

Perimeter of a rectangle 1

1 Liam draws a rectangle. He uses cubes to measure the length of each side. Each cube is 1 cm long.

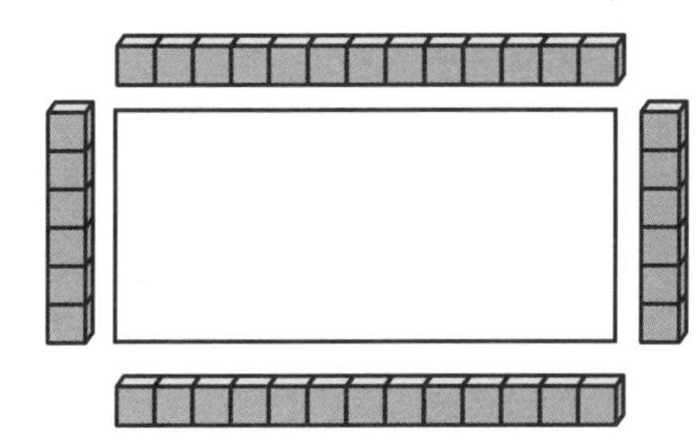

What is the perimeter of the rectangle?

☐ + ☐ + ☐ + ☐ = ☐ cm

2 Find the perimeters of these rectangles.

A

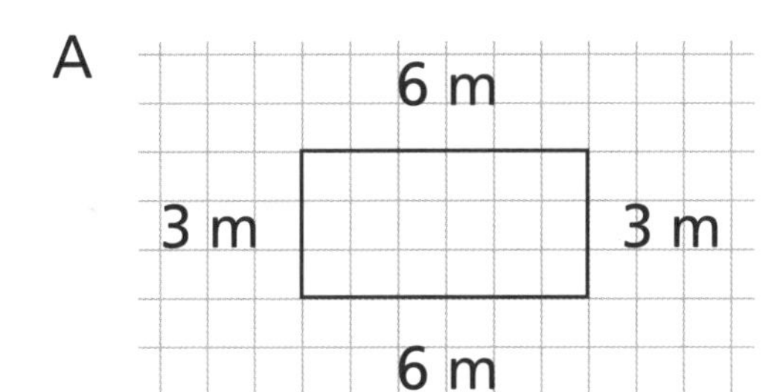

C

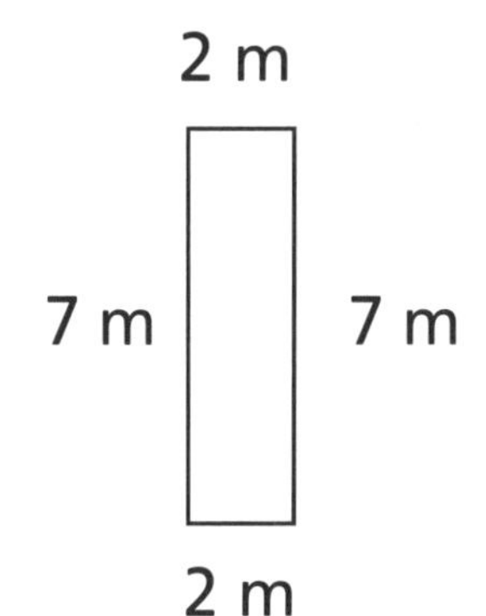

B

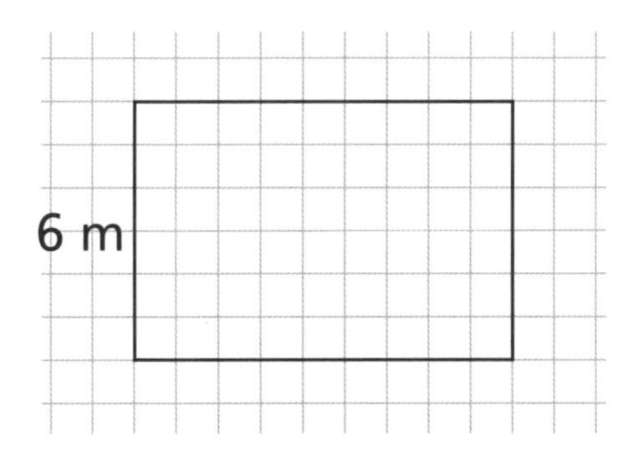

D

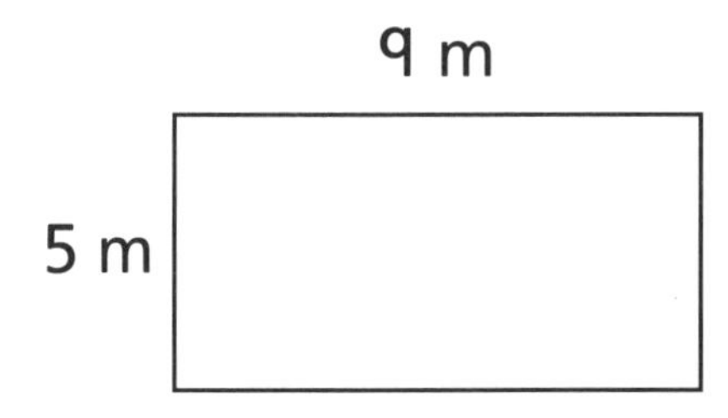

Rectangle	Perimeter
A	☐ m
B	☐ m
C	☐ m
D	☐ m

3 Tick all children who are showing perimeter.

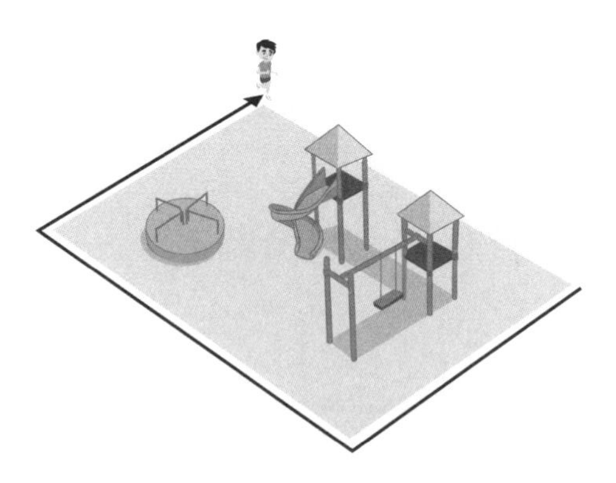

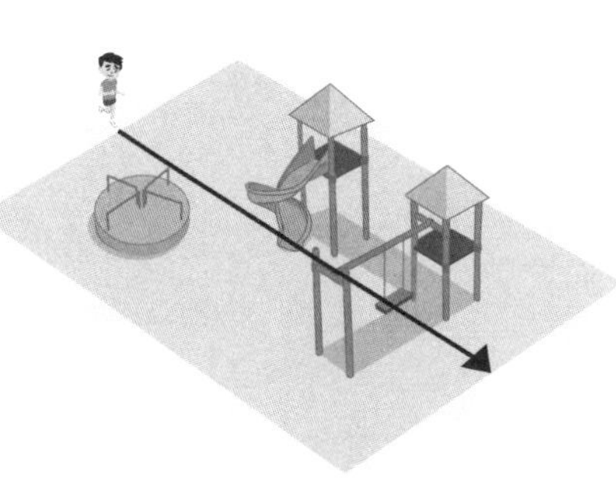

4 Each square has a length of 5 m.

a) Label the length and the width of this swimming pool.

m

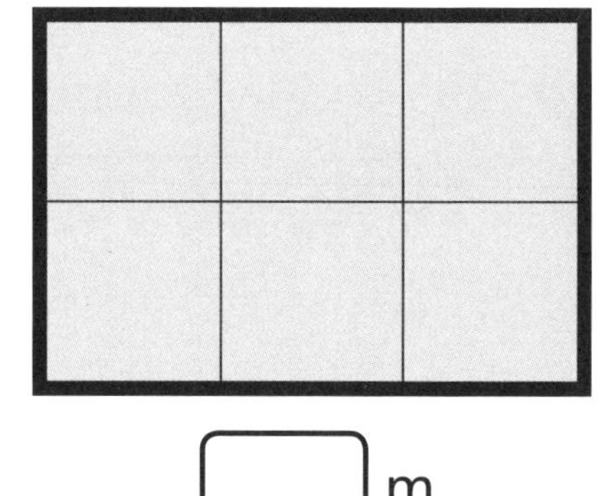

m

b) What is its perimeter?  m

5 The school field is 50 m long and 23 m wide. 

Jack runs the length of the field 3 times.

Sam runs around the perimeter once (1 time).

Who has run further?

_______________ has run further.

Explain your answer.

____

____

6 A carpet company sells square rugs. Two sizes are shown.

CHALLENGE

5 m

6 m

a) Complete the table.

Side length	5 m	6 m	7 m	☐ m	☐ m
Perimeter	☐ m	☐ m	☐ m	32 m	40 m

b) What do you notice about the perimeters of the rugs?
Why is this?

____

____

____

____

Reflect

A classroom is a rectangle. Its length is 6 m. Its width is 5 m.

To work out the perimeter of this classroom, I would …

____

____

→ Textbook 4A p160

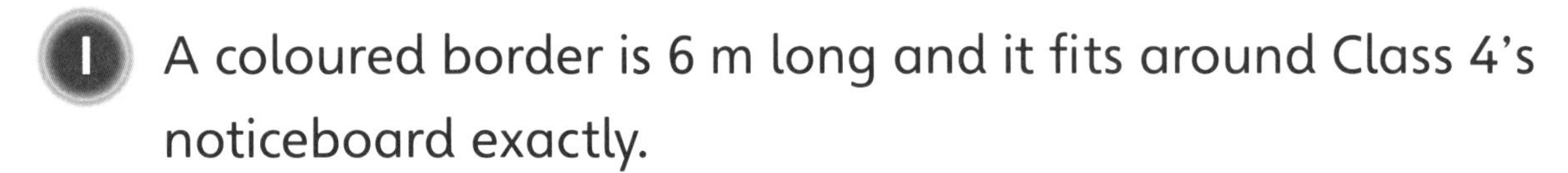

Perimeter of a rectangle 2

1 A coloured border is 6 m long and it fits around Class 4's noticeboard exactly.

The height of the noticeboard is 1 m.

What is its length?

Complete the bar model to help.

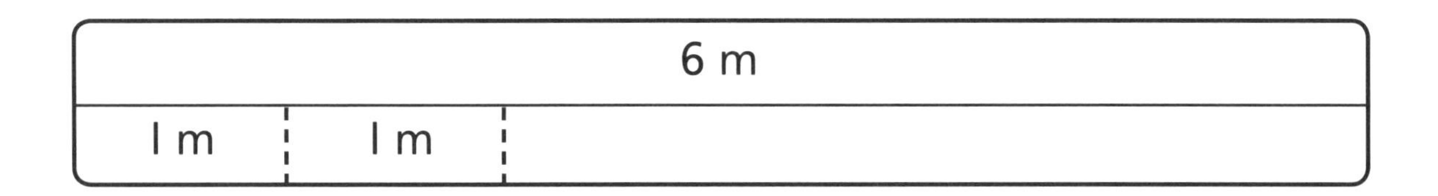

☐ ÷ 2 = ☐

The length of the noticeboard is ☐ m.

2 Tim has 40 sticks.

He uses all of them to make the outline of a rectangle that is 12 sticks long.

What are the perimeter, length and width of the rectangle?

Perimeter = ☐ sticks

Length = ☐ sticks

Width = ☐ sticks

3 Draw lines to match each rectangle with its missing measurement.

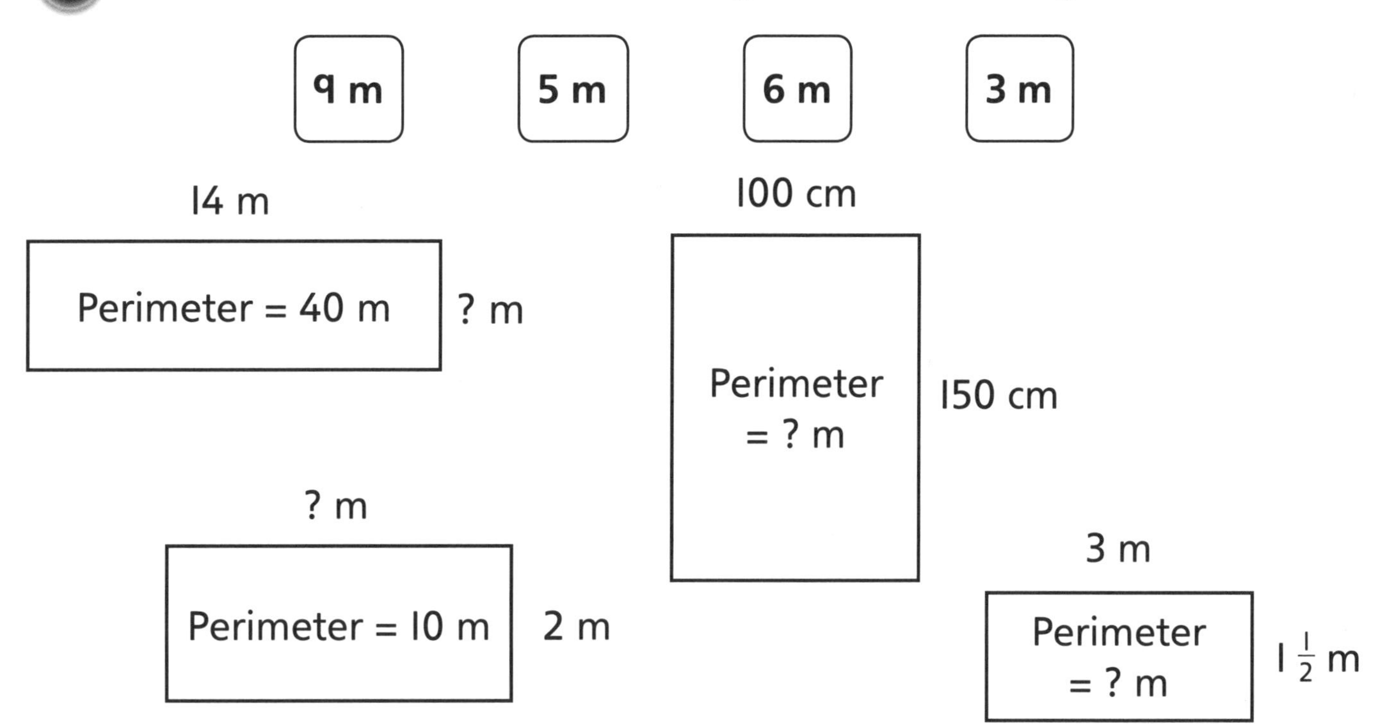

4 The perimeter of a rectangle is 16 cm.

a) Complete the table to show the different rectangles it could be.

Width	Length
1 cm	7 cm
2 cm	☐ cm
3 cm	☐ cm
4 cm	☐ cm

b) What do you notice about the last shape you listed?

____

CHALLENGE

5 This square table has side lengths of 70 cm.

a) What is its perimeter? [ ] cm

b) Two tables are put next to each other.

What is the perimeter now? [ ] cm

Draw a diagram to show your answer.

70 cm

Reflect

The perimeter of a rectangle is 12 cm. Its width is 1 cm.

Explain what to do to find out its length.

→ Textbook 4A p164

Perimeter of rectilinear shapes 

1 A gardener uses wooden edging around a flower bed.

Each piece of edging is 1 m long.

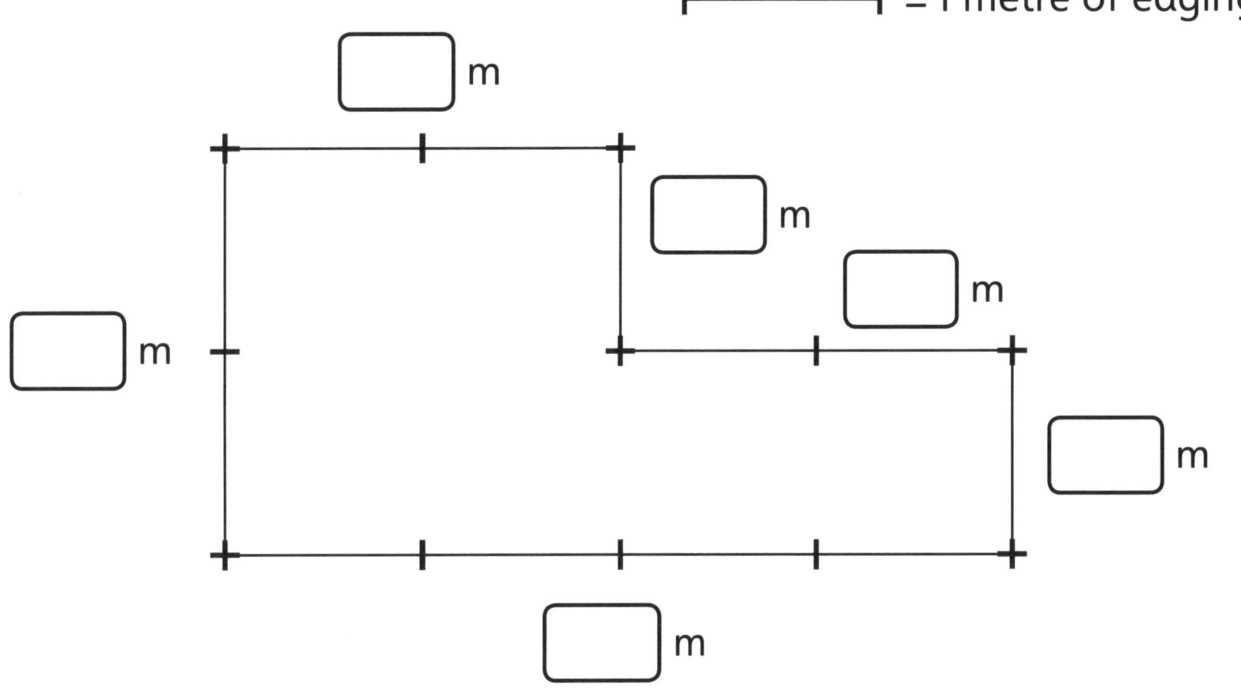

a) Complete the measurements of each side.

b) Work out the perimeter of the flower bed.

The perimeter of the flower bed is [ ] m.

2 Label each shape with its perimeter.

a)

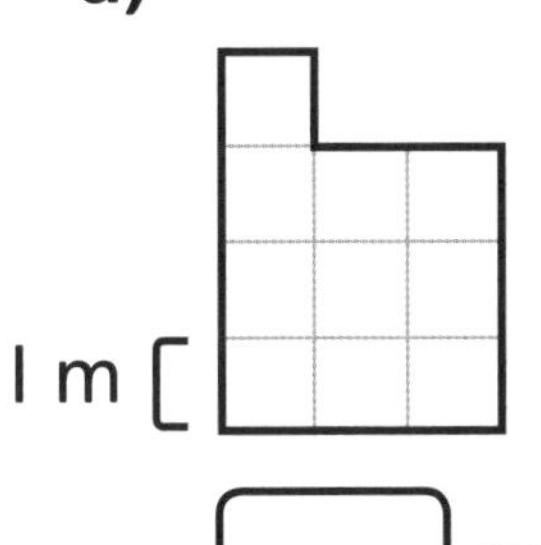

[ ] m

b)

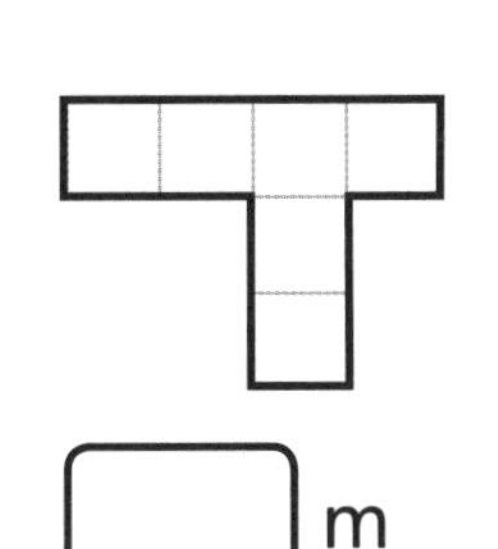

[ ] m

c)

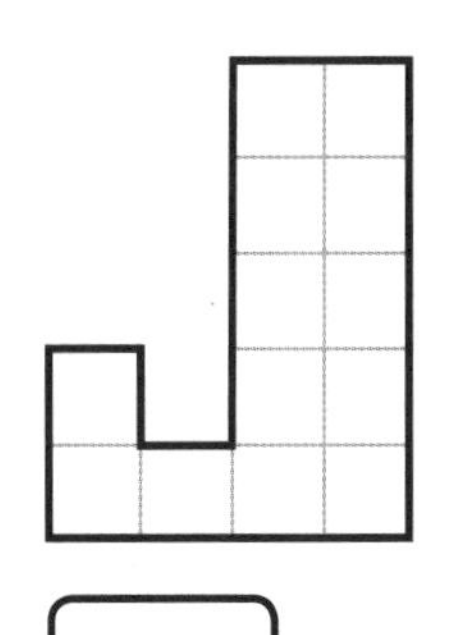

[ ] m

d)

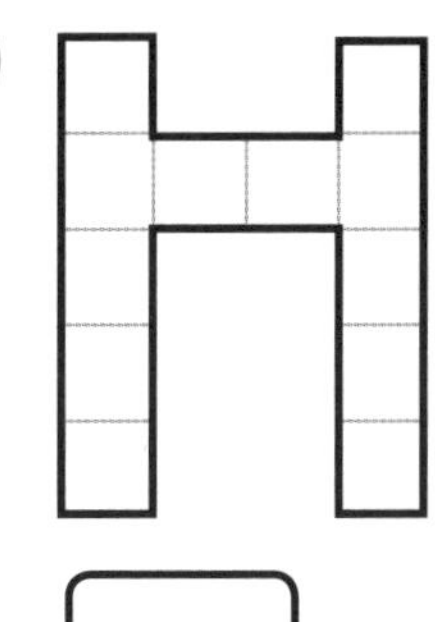

[ ] m

3 Lottie is designing a badge in the shape of the letter L.

The lengths of its sides are: 7 cm, 2 cm, 4 cm, 3 cm, 3 cm and 5 cm.

a) The perimeter of the badge is

☐ cm.

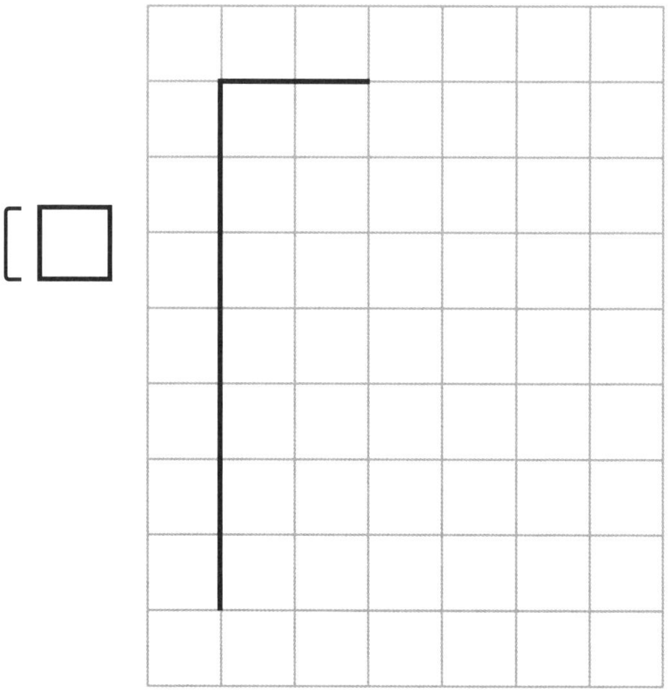

I cm

b) Use the measurements to draw the badge. The first two lines are done for you.

4 The sides of this rectilinear shape, in order, are 3 m, I m, 4 m, I m, 5 m, I m, 2 m, I m.

Label the diagram and find the perimeter.

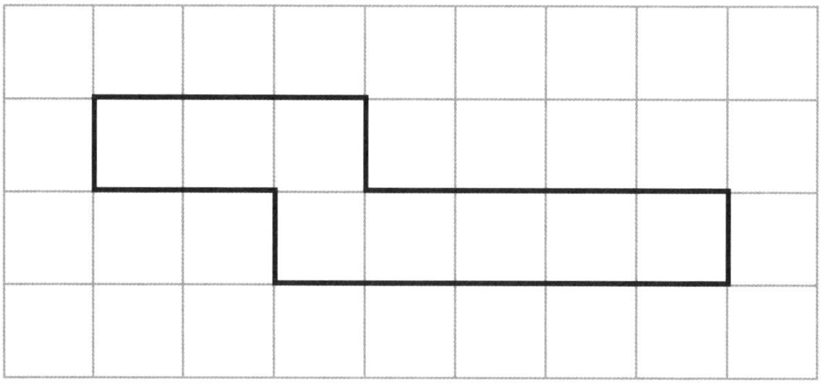

5 **a)** Use the squared paper below to draw a rectilinear outline of a factory with two chimneys.

Keep to the squares. Your factory needs to have a flat roof.

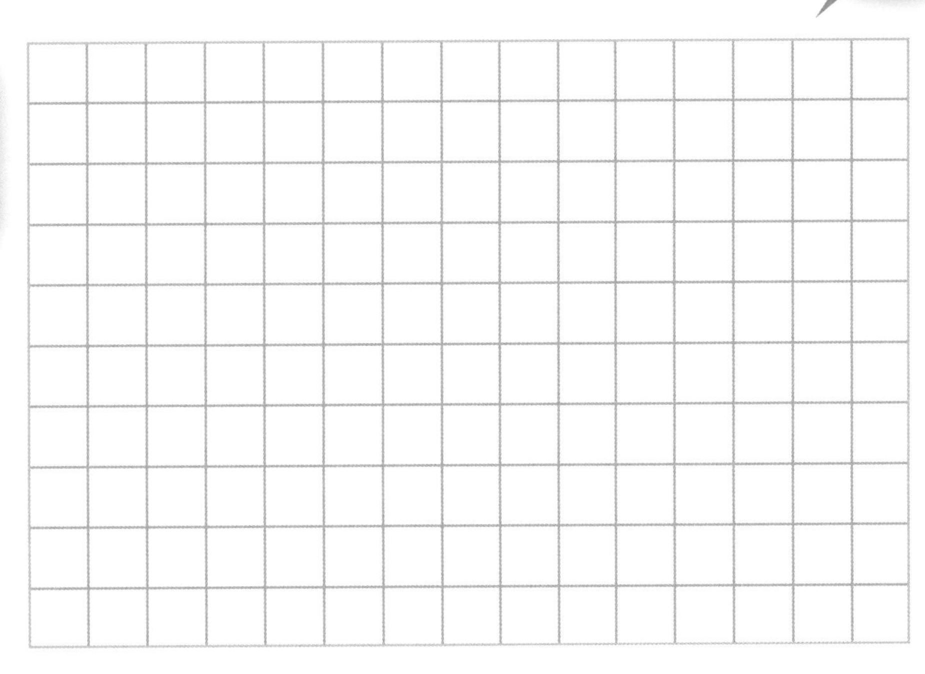

b) Label the length of each side.

c) What is the perimeter of the factory? ______________

Reflect

Amy is working out the perimeter of this shape.

1 cm [□

1	2	3	4	5			
26				6			
25				7			
24				8			
23				9	10	11	12
22							13
21	20	19	18	17	16	15	14

She says that it is 26 cm.

Is Amy correct? Yes / No (Circle your choice.) Explain why.

____

→ Textbook 4A p168

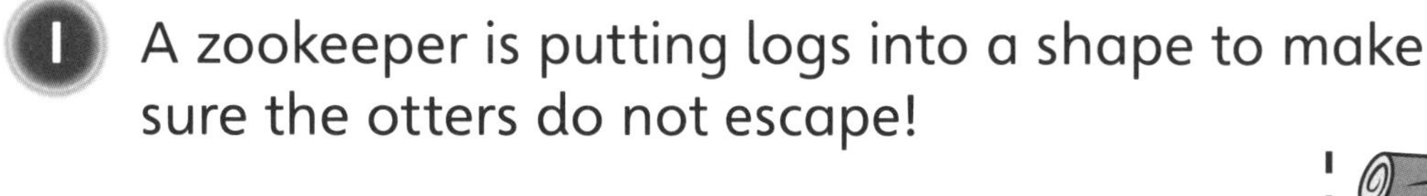

Perimeter of rectilinear shapes 2

1 A zookeeper is putting logs into a shape to make sure the otters do not escape!

a) He has nearly finished.

How many more logs does he need to finish the shape?

He needs ☐ more logs.

b) What is the perimeter of the shape?

The perimeter is ☐ logs.

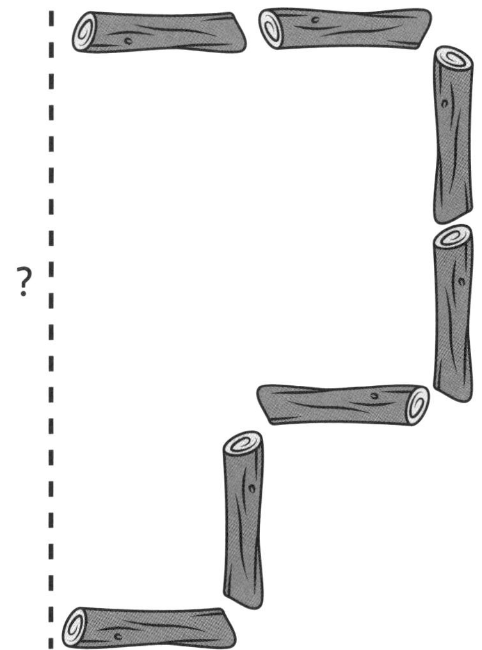

2 Complete the sentences.

Side A is equal to the total of

☐ cm and ☐ cm.

It is ☐ cm long.

Side B is equal to ☐ cm less than

☐ cm.

It is ☐ cm long.

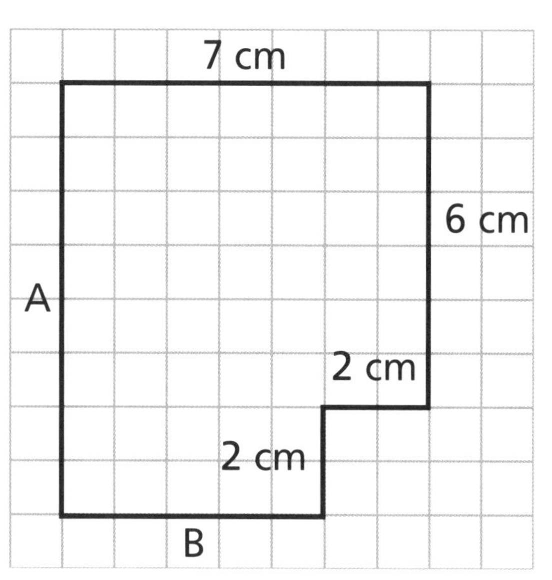

The perimeter of this shape is ☐ cm.

3 Find the measurement of each missing side, then calculate the perimeters of the two shapes.

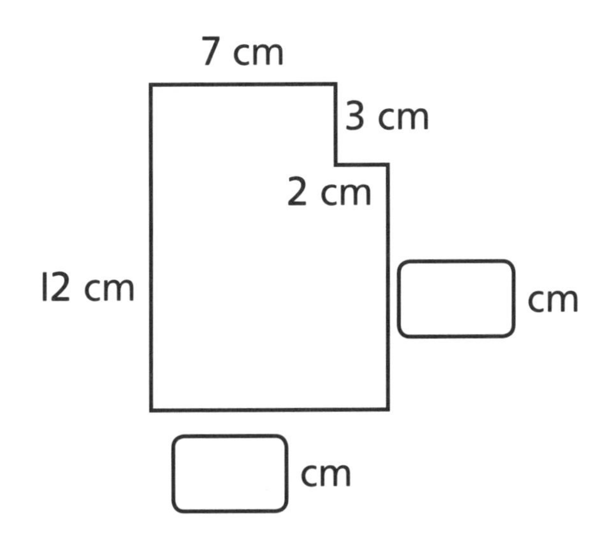

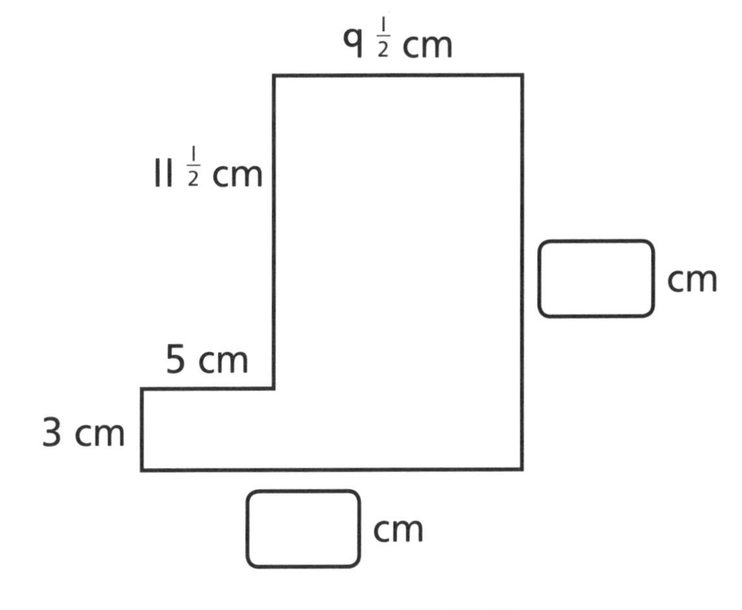

perimeter = ☐ cm

perimeter = ☐ cm

4 This rectilinear shape does not have all its measurements marked.

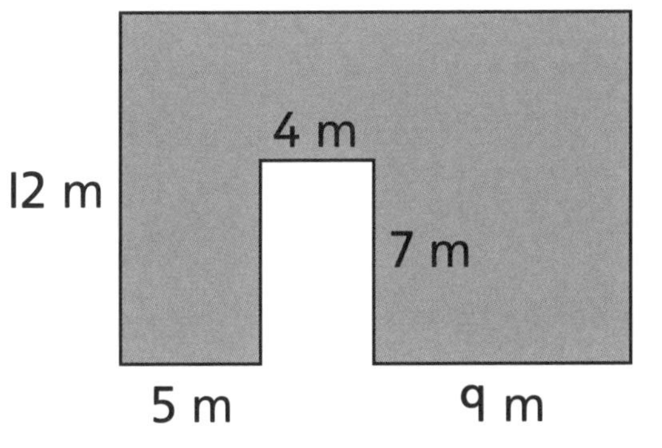

a) Explain how to work out the missing values.

____

____

b) Work out the perimeter.

____

____

The perimeter is ☐ m.

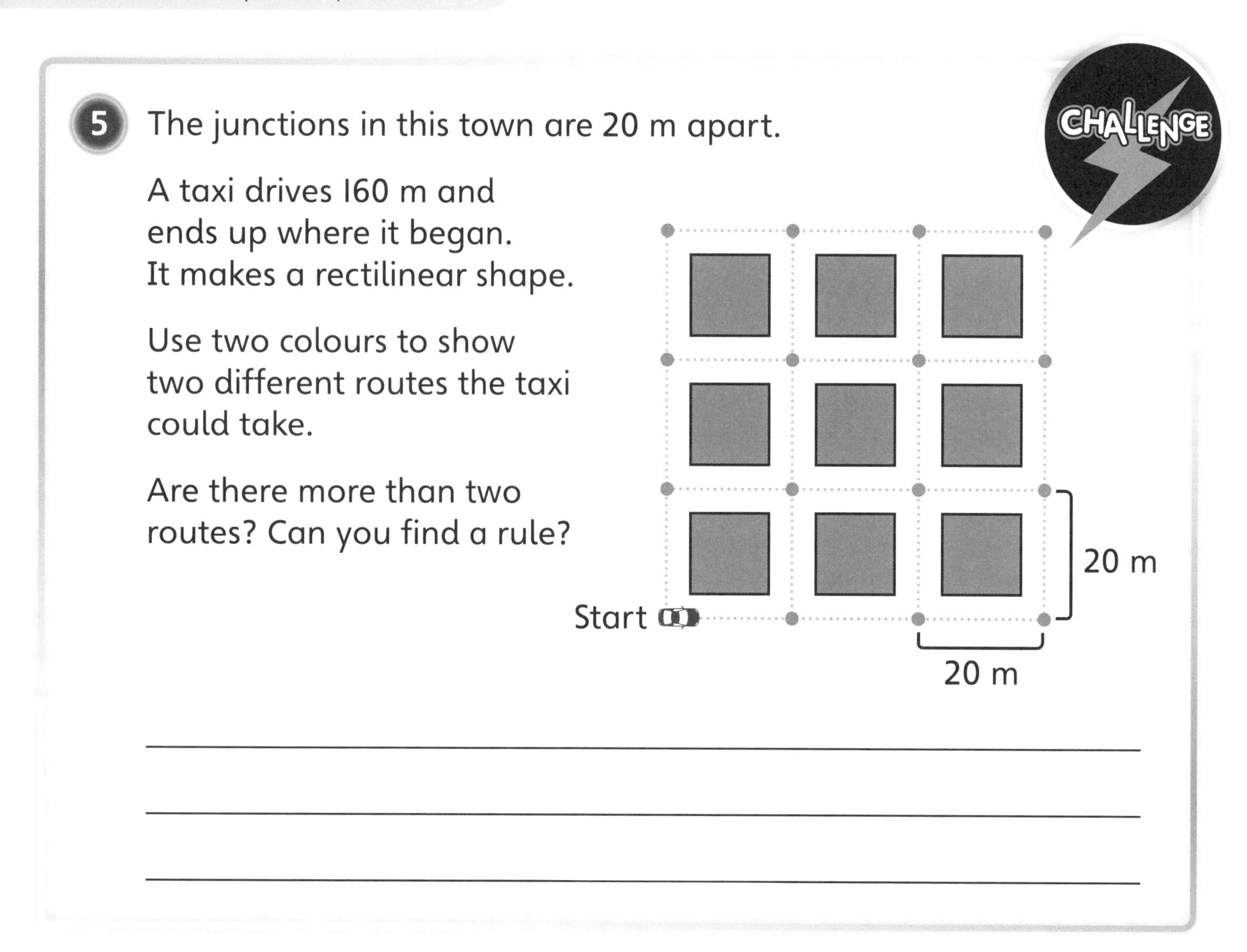

5 The junctions in this town are 20 m apart.

A taxi drives 160 m and ends up where it began. It makes a rectilinear shape.

Use two colours to show two different routes the taxi could take.

Are there more than two routes? Can you find a rule?

____

____

____

Reflect

Design a rectilinear perimeter problem for a friend to answer.

Explain how you expect your friend to work out the perimeter.

____

____

____

→ Textbook 4A p172

End of unit check

My journal

1. Draw two different rectilinear shapes, each with a perimeter of 18 cm.

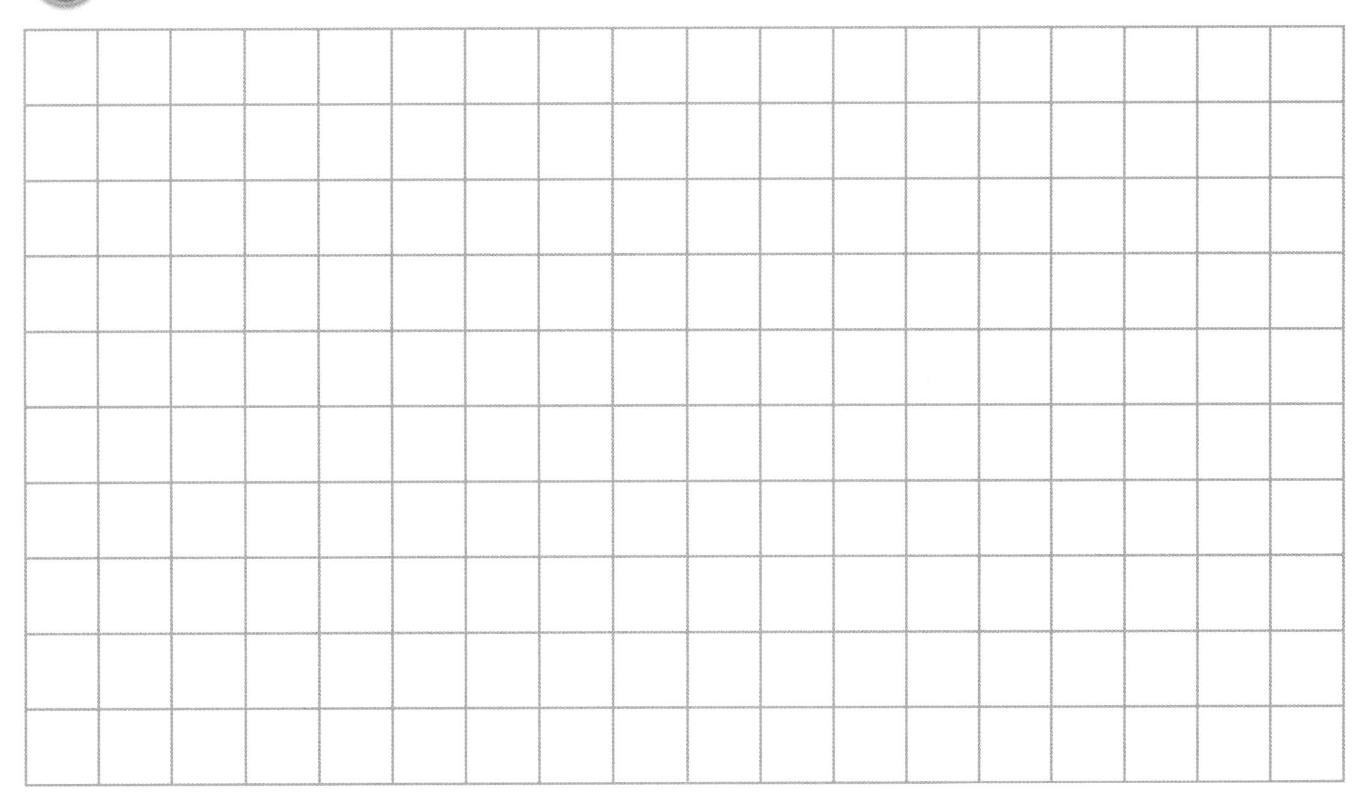

Explain how you decided on the measurements for the two shapes.

____

____

____

Power check

How do you feel about your work in this unit?

Power play

This shape has a perimeter of 12 sticks.

Remove 1 stick and try to make a new shape with a perimeter of 11 sticks.

Ensure the shape is rectilinear (straight sides and right angles at all corners).

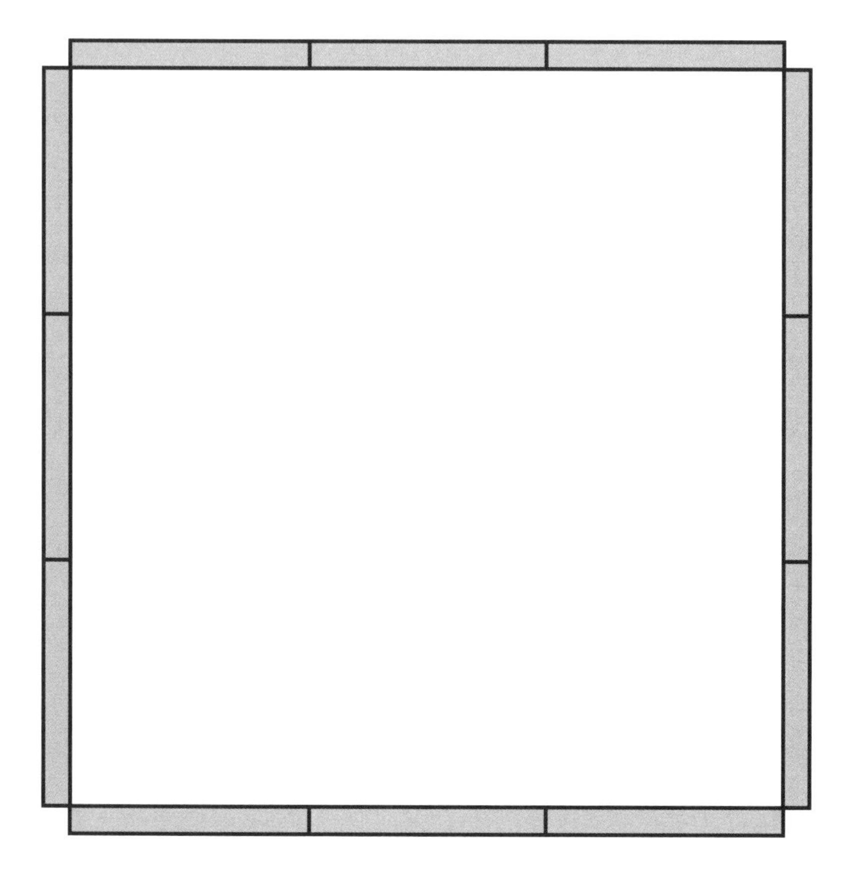

Remove 1 stick each time.

Which perimeters can you make a shape for?

Which perimeters are impossible? Why is this?

Try starting with 20 sticks and **adding** one more stick each time. What do you notice?

→ Textbook 4A p176

Multiplying by multiples of 10 and 100

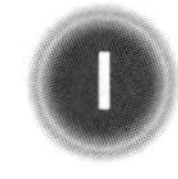

1

a) How many boxes of pencils are there?

☐ × 5 = ☐

There are ☐ boxes of pencils.

b) How many pencils are there in total?

☐ × ☐ = ☐

There are ☐ pencils in total.

2

a) How many jars of sweets are there?

☐ × ☐ = ☐

There are ☐ jars of sweets.

b) How many sweets are there in total?

☐ × ☐ = ☐

There are ☐ sweets in total.

3 Work out 7×30 using three different methods.

Method 1:

$7 \times 30 =$

☐ + ☐ + ☐ + ☐ + ☐ + ☐ + ☐ = ☐

Method 2:

7×3 ones = ☐ ones = ☐

So, 7×3 tens = ☐ tens = ☐

Method 3:

$7 \times 3 =$ ☐

Then, ☐ $\times 10 =$ ☐

4 What calculations are shown on the number lines?

a)

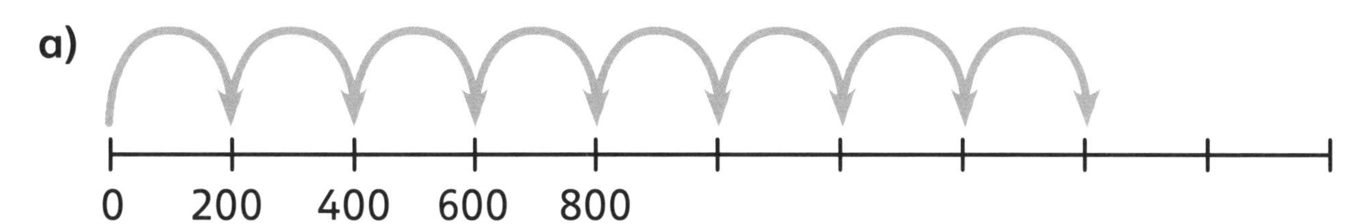

☐ × ☐ = ☐

b)

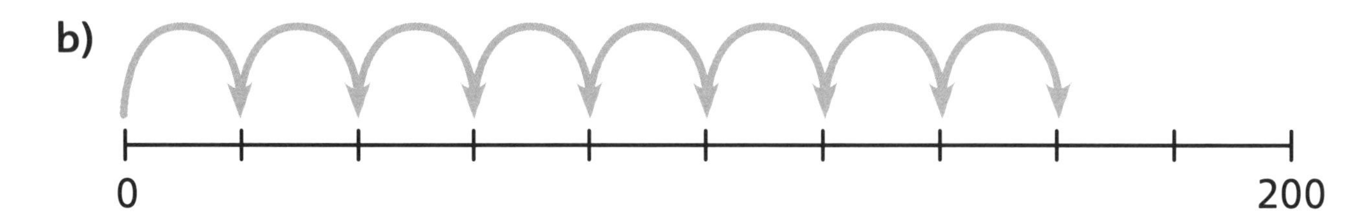

☐ × ☐ = ☐

5 Find the solutions to these calculations.

a) $7 \times 4 = \square$

$7 \times 40 = \square$

$7 \times 400 = \square$

b) $8 \times 30 = \square$

$8 \times 300 = \square$

$3 \times 8 = \square$

c) $9 \times 2 = \square$

$9 \times 20 = \square$

$200 \times 9 = \square$

d) $9 \times 50 = \square$

$80 \times 9 = \square$

$600 \times 4 = \square$

6 Work out the missing number to make the calculation correct.

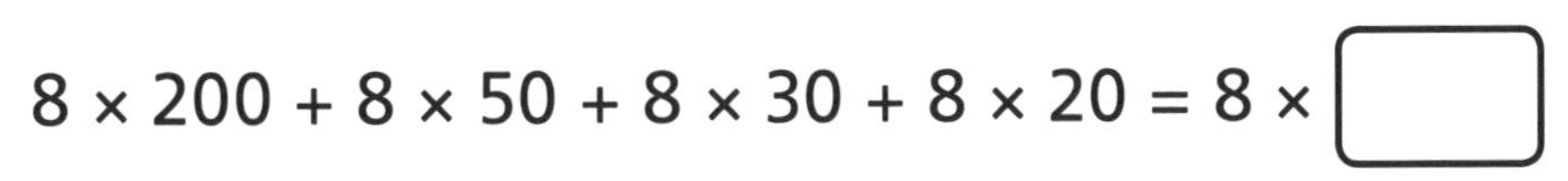

$8 \times 200 + 8 \times 50 + 8 \times 30 + 8 \times 20 = 8 \times \square$

Explain your method.

____

____

Reflect

Choose one of these calculations and explain how you can use $7 \times 4 = 28$ to work it out.

7×40 70×4 700×4 7×400

____

____

→ Textbook 4A p180

Dividing multiples of 10 and 100

1 Work out the following calculations.

a)

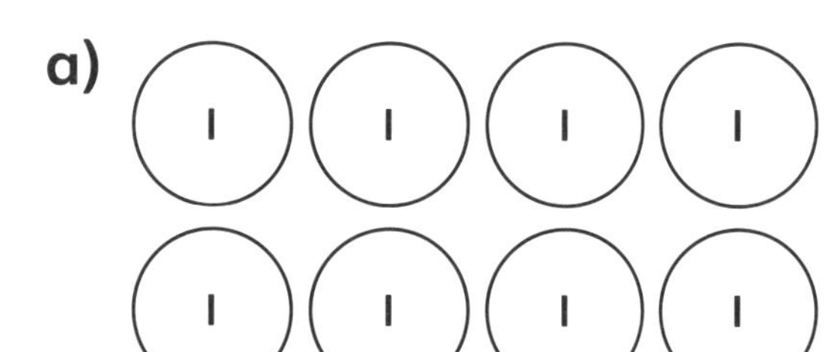

$8 \div 4 = \square$

b)

$80 \div 4 = \square$

c)

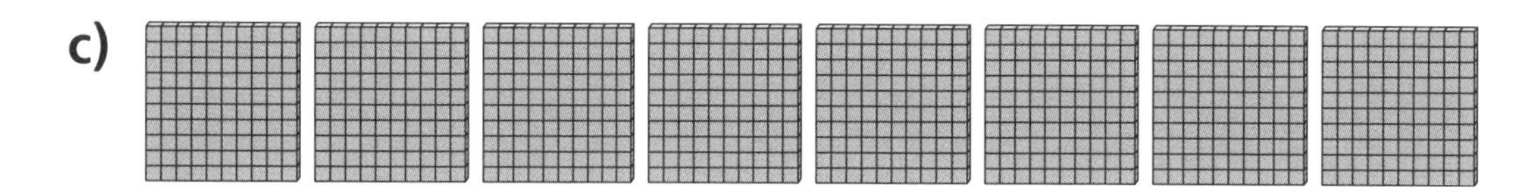

$800 \div 4 = \square$

2 What division calculation does each number line show?

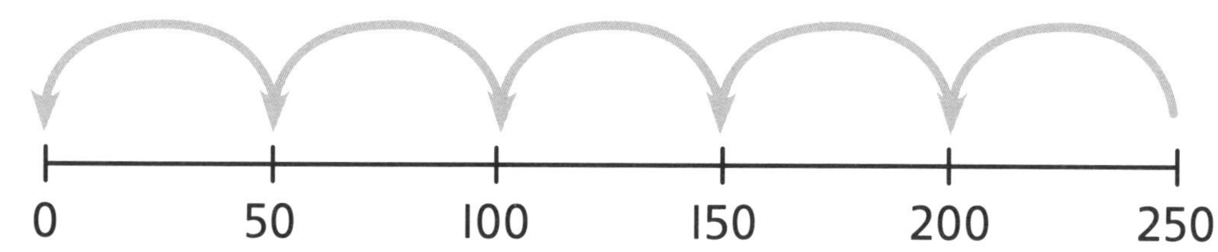

a) $\square \div \square = \square$

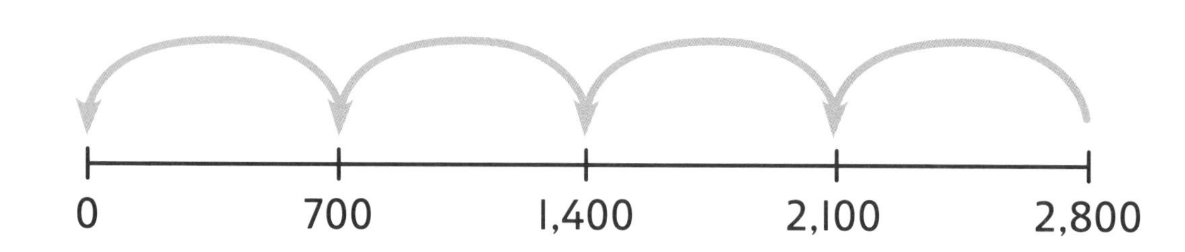

b) $\square \div \square = \square$

3 Draw lines to match each calculation to the correct answer.

400 ÷ 5		1,600 ÷ 2
480 ÷ 6	8	4 thousands ÷ 5
40 ÷ 5	80	720 ÷ 9
32 ÷ 4	800	800 ÷ 10
32 tens ÷ 4		

4 Find the solutions to these calculations.

a) 12 ÷ 2 = ☐

120 ÷ 2 = ☐

1,200 ÷ 2 = ☐

b) 18 ones ÷ 3 = ☐

18 tens ÷ 3 = ☐

18 hundreds ÷ 3 = ☐

c) 33 ÷ 3 = ☐

330 ÷ 3 = ☐

3,300 ÷ 3 = ☐

d) 350 ÷ 5 = ☐

320 ÷ 2 = ☐

80 ÷ 4 = ☐

1,600 ÷ 4 = ☐

5 Here is a function machine.

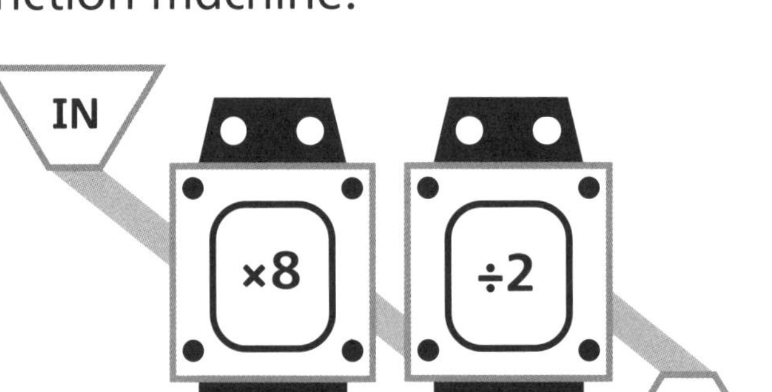

Complete the table.

IN	6	60	600	20	300		
OUT						50	120

What do you notice?

____

____

Reflect

Explain how to divide 1,200 by 4 and discuss it with a friend.

Did your friend use the same method?

- ______________________________
- ______________________________
- ______________________________
-

Multiplying by 0 and 1

1 How many muffins are on each group of plates?

Draw lines to match each picture to the correct multiplication sentence.

a)

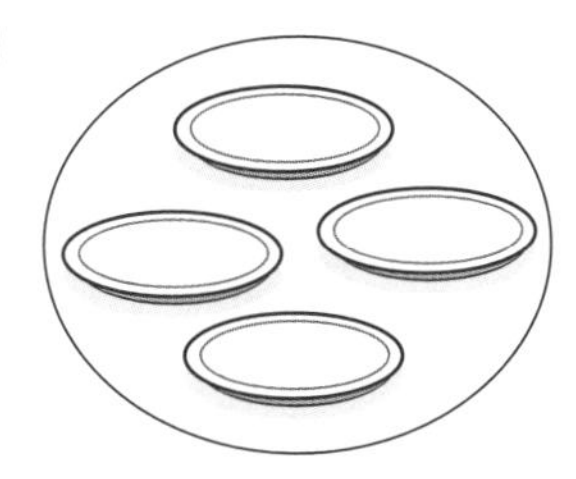

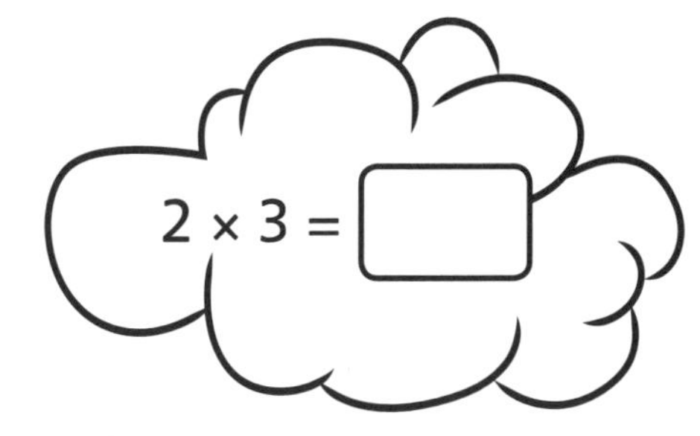

b)

c)

d)

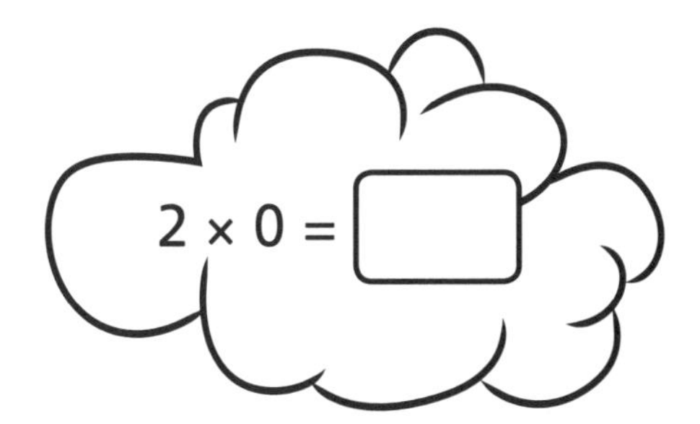

e)

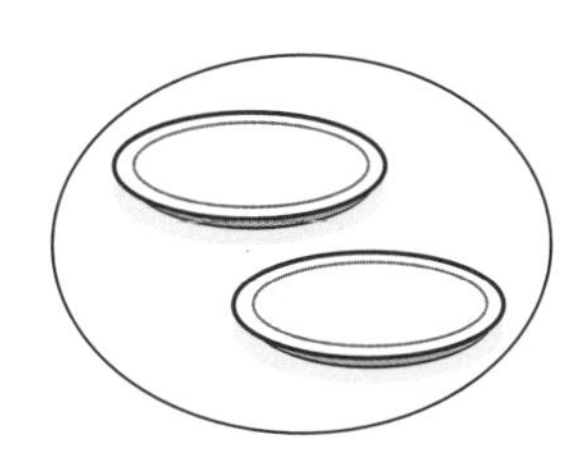

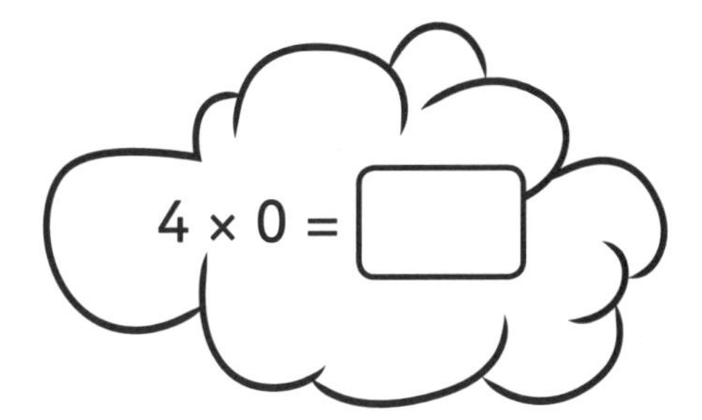

2 Here are 4 trays. Each tray contains the same amount of equipment.

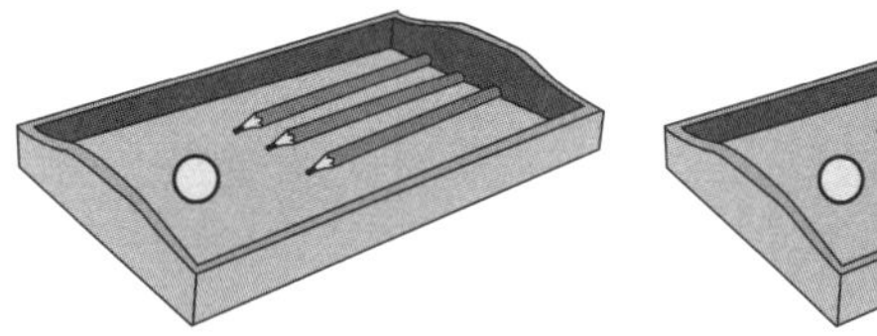 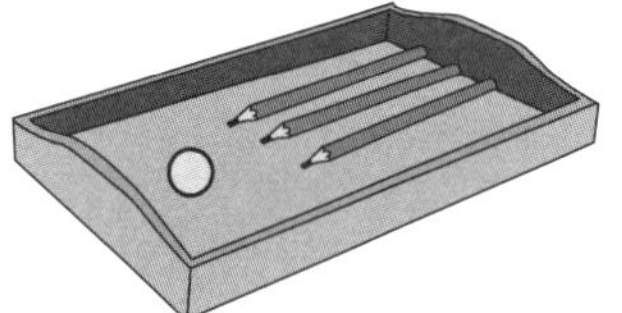 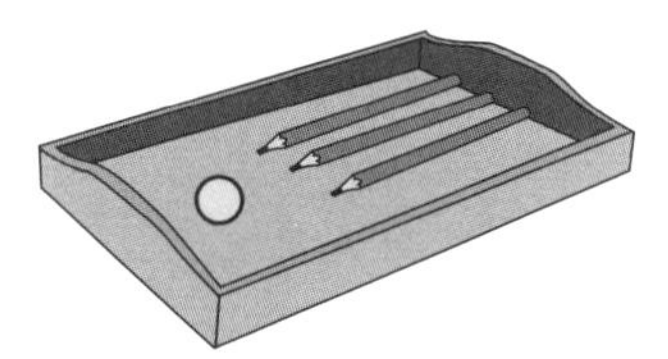 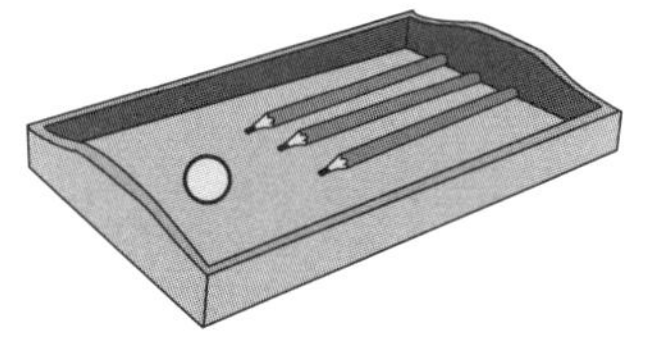

a) How many counters are there in total?

☐ × ☐ = ☐ There are ☐ counters in total.

b) How many pencils are there in total?

☐ × ☐ = ☐ There are ☐ pencils in total.

c) How many cubes are there in total?

☐ × ☐ = ☐ There are ☐ cubes in total.

3 Circle the multiplications that have an answer of 0.

a) 3 × 0 **c)** 0 × 10 **e)** 15 × 0 **g)** 6 × 1

b) 1 × 5 **d)** 3 × 8 **f)** 0 × 5 **h)** 1 × 0

What is the same about all of the calculations you have circled?

___

4 Fill in the missing numbers to make the calculations correct.

a) $7 \times 0 = \square$

b) $1 \times 9 = \square$

c) $\square \times 1 = 15$

d) $\square = 127 \times 0$

5 Kate has a function machine.

Kate puts the number 5 into the function machine.

What number does Kate get out? $\square$

IN → ×2 → ×3 → ×4 → ×1 → ×0 → OUT

Reflect

$\square \times 0 = \square$

$\square \times 1 = \square$

Look at the two calculations above. What can you say about the numbers that go in each of the boxes?

→ Textbook 4A p188

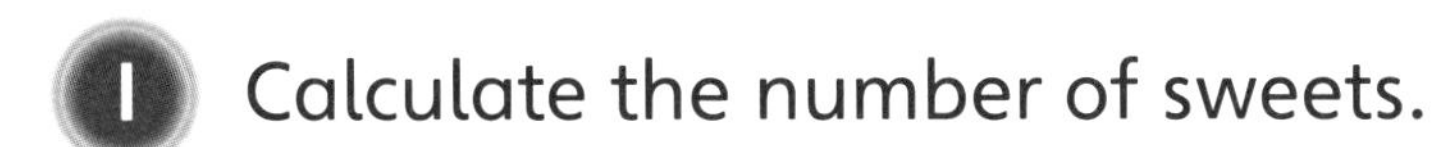

Dividing by 1

1 Calculate the number of sweets.

a) There are 6 sweets. The sweets are shared equally between 1 person.

How many sweets does the person receive?

☐ ÷ ☐ = ☐

The person receives ☐ sweets.

b) There are 6 sweets. The sweets are shared between 6 people.

How many sweets does each person receive?

☐ ÷ ☐ = ☐

Each person receives ☐ sweet.

2 What mistake has Amelia made?

____

____

____

3 Circle the calculations that have an answer of 1.

$8 \div 8$ $8 \div 1$ $5 \div 5$ $16 \div 16$

$20 \div 2$ $7 \div 7$ $2 \div 1$ $150 \div 150$

4 **a)** Find the solutions to these calculations.

$3 \div 1 =$ ☐ $4 \div 1 =$ ☐ $5 \div 1 =$ ☐

$10 \div 1 =$ ☐ $14 \div 1 =$ ☐ $20 \div 1 =$ ☐

Use the calculations to complete the following sentence.

When you divide a number by 1 ______________________________

____

____

____

b) Find the solutions to these calculations.

$3 \div 3 =$ ☐ $4 \div 4 =$ ☐ $5 \div 5 =$ ☐

$10 \div 10 =$ ☐ $14 \div 14 =$ ☐ $20 \div 20 =$ ☐

Use the calculations to complete the following sentence.

When you divide a number by itself ______________________________

____

____

____

5 Fill in the missing numbers to make the calculations correct.

a) $11 \div 1 = \square$

b) $11 \div 11 = \square$

c) $\square = 25 \div 25$

d) $9 \div \square = 9$

e) $12 \div \square = 1$

f) $\square \div 1 = 70$

g) $\square \div 1 = 0$

h) $8 \div \square = 7 \div 7$

6 The square and the pentagon represent numbers. Look at the number sentence then tick the correct statement.

$\square \div 1 > ⬠ \div 1$

CHALLENGE

The square is equal to the pentagon. ☐

The square is greater than the pentagon. ☐

The pentagon is greater than the square. ☐

Explain your answer.

____

____

Reflect

$\square \div \square = 1$

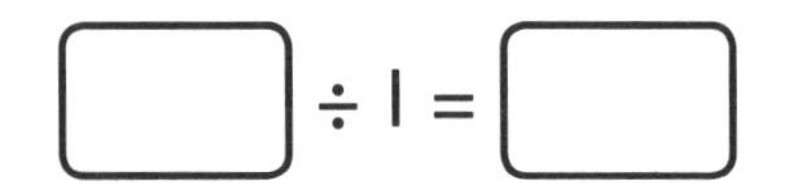

$\square \div 1 = \square$

Look at the two calculations above. What can you say about the numbers that go in each of the boxes?

____

____

→ Textbook 4A p192

Multiplying and dividing by 6

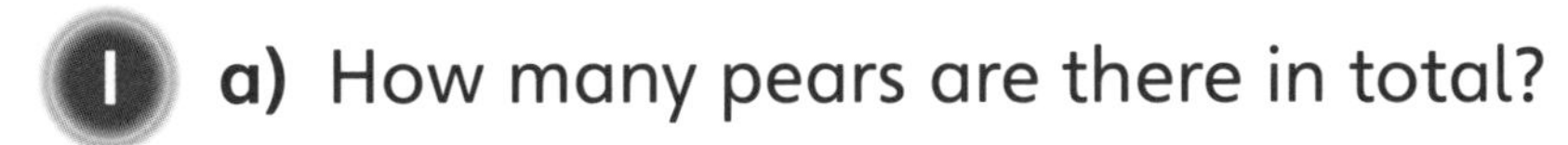

1 **a)** How many pears are there in total?

☐ × ☐ = ☐

There are ☐ pears in total.

b) How many flowers are there in total?

☐ × ☐ = ☐

There are ☐ flowers in total.

2 Alex has 48 triangles.

6 triangles are put together to make a hexagon.

How many hexagons can Alex make in total?

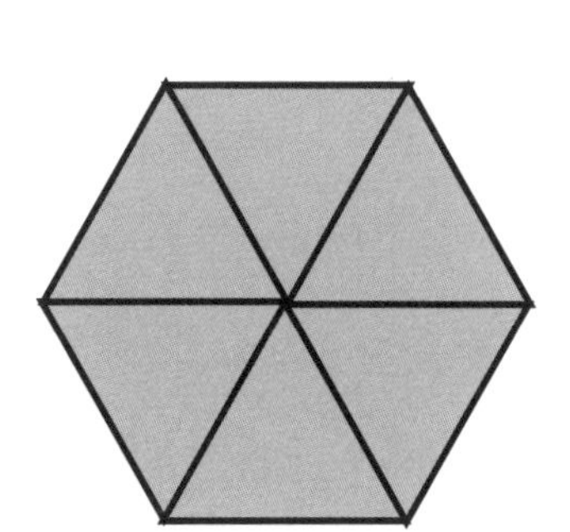

☐ ÷ ☐ = ☐

Alex can make ☐ hexagons.

3 £24 is shared between 6 people.

How much does each person receive?

☐ ○ ☐ ○ ☐

Each person receives £☐.

4 A group of 6 people are flying from the UK to France and then to Canada.

a) Each flight to France costs £90.

How much do the flights to France cost in total for the group?

☐ ○ ☐ = ☐

The total cost of the flights to France is £☐.

b) The flights from France to Canada cost £1,800 in total.

How much does each person's flight cost?

☐ ○ ☐ = ☐

Each flight to Canada costs £☐.

c) How much does each person pay in total for their flights?

☐ ○ ☐ = ☐

The total cost for each person is £☐.

CHALLENGE

5 A rectangle has a length of 12 cm and width of 6 cm.

6 cm

12 cm

6 rectangles are used to make a longer rectangle.

What is the perimeter of the new shape above?

The perimeter of a shape is the length all the way around the outside.

The perimeter of the new shape is ☐ cm.

Reflect

Draw or write your own story involving multiplication or division by 6. Ask your partner to find the solution to it.

→ Textbook 4A p196

6 times-table

1 Which 6 times-table facts do these pictures show?

a)

☐ × ☐ = ☐

b)

☐ × ☐ = ☐

2 Find the solution to these calculations.

a) 3 × 6 = ☐

b) 1 × 6 = ☐

c) 6 × 6 = ☐

d) 12 × 6 = ☐

e) ☐ = 6 × 10

f) 0 = 6 × ☐

g) ☐ × 6 = 24

h) 9 × 6 = ☐

i) 6 ÷ 6 = ☐

j) 24 ÷ 6 = ☐

k) 42 ÷ 6 = ☐

l) ☐ = 66 ÷ 6

m) ☐ ÷ 6 = 0

n) ☐ ÷ 6 = 1

o) ☐ ÷ 10 = 6

3 Fill in the missing numbers.

a)

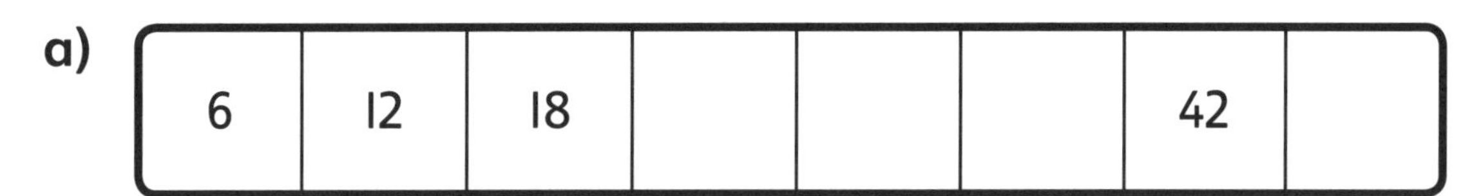

6	12	18				42	

b)

60				36		

c) 0, 60, 120, ______, ______, ______, ______

4 Circle all of the numbers in the 6 times-table.

60	6	0	2	1	15	28	200	120	126

5 $12 \times 6 = 72$

Use this to work out 13×6.

$13 \times 6 =$ ☐

6 Complete each calculation using <, > or =

a) 2×6 ◯ 10

b) $36 \div 6$ ◯ 30

c) 5×6 ◯ 7×6

d) $18 \div 6$ ◯ $24 \div 6$

e) 9×6 ◯ 6×9

f) 15×6 ◯ 6×12

7 Complete the number sentences.

a) 4 × 6 = ☐

40 × 6 = ☐

400 × 6 = ☐

6 × 40 = ☐

b) 900 × 6 = ☐

6 × 70 = ☐

50 × 6 = ☐

☐ × 6 = 1,200

8 a) How can you use the answer to 8 × 3 to work out 8 × 6?

____

____

b) How can you use the answer to 8 × 5 to work out 8 × 6?

____

____

Reflect

How fast can you complete the 6 times-table?

0 × 6 = ☐	1 × 6 = ☐	2 × 6 = ☐	3 × 6 = ☐
4 × 6 = ☐	5 × 6 = ☐	6 × 6 = ☐	7 × 6 = ☐
8 × 6 = ☐	9 × 6 = ☐	10 × 6 = ☐	11 × 6 = ☐
12 × 6 = ☐	Time taken: ________		

Circle the answers you knew without having to work them out.

→ Textbook 4A p200

Multiplying and dividing by 9

1 **a)** How many hearts are there in total?

☐ × ☐ = ☐ There are ☐ hearts.

b) How many spades are there in total?

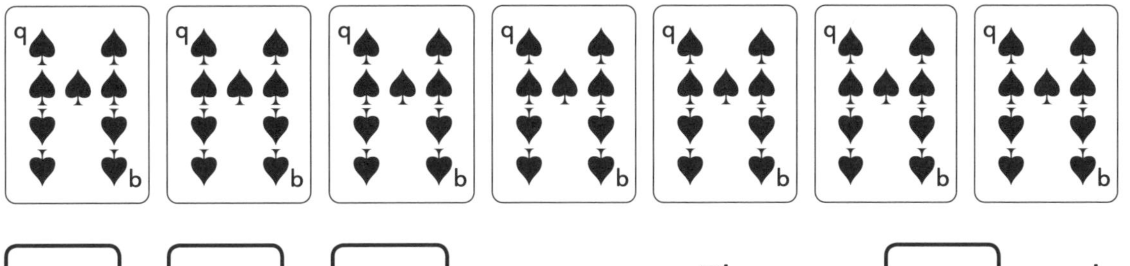

☐ × ☐ = ☐ There are ☐ spades.

2 Circle groups of triangles to show how you can use the array to work out 18 ÷ 9.

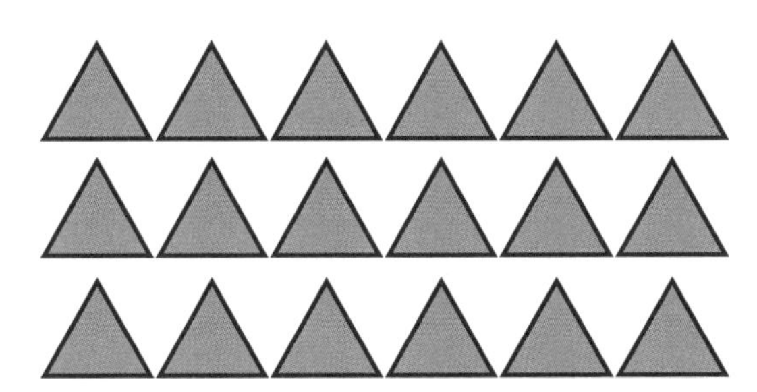

18 ÷ 9 = ☐

____

____

____

Did you use grouping or sharing?

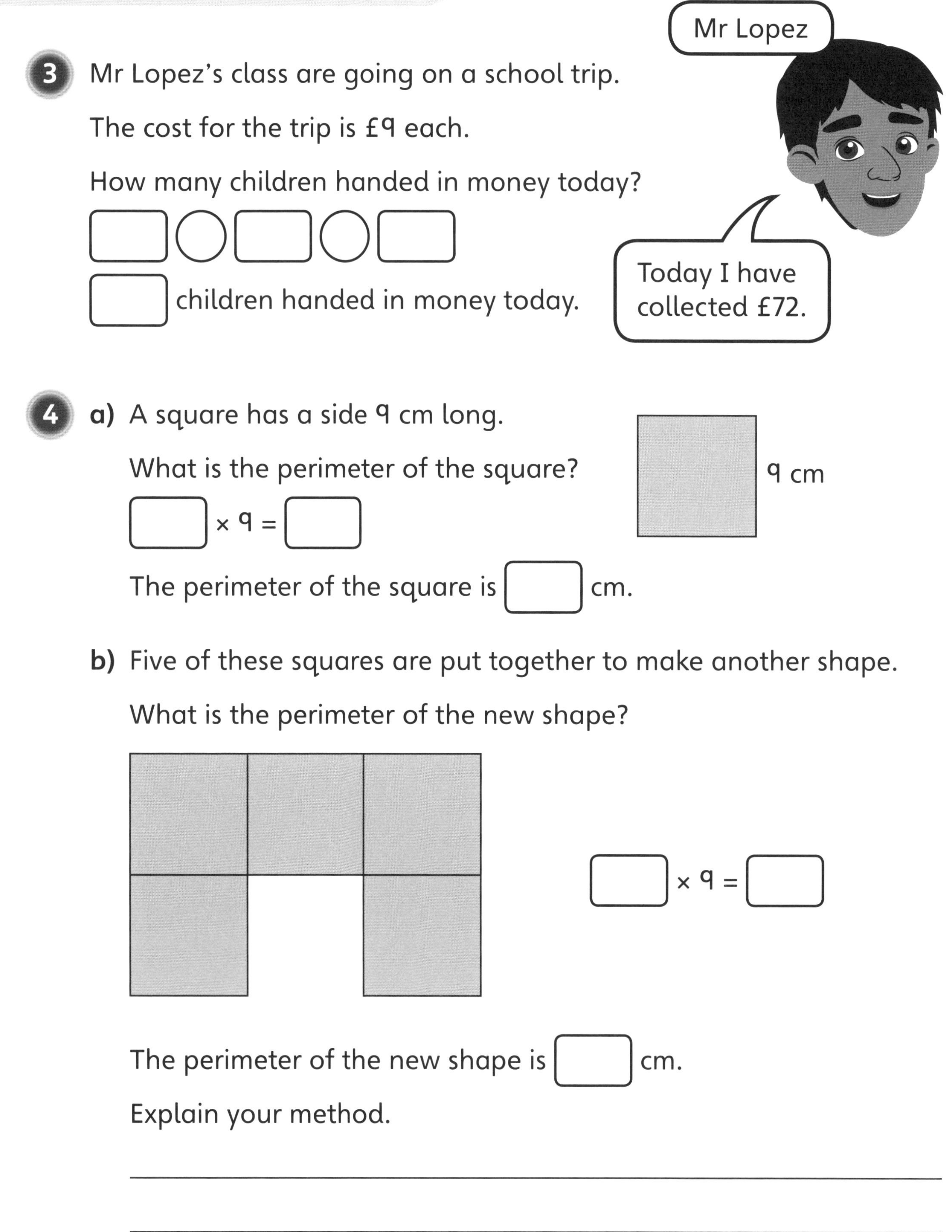

3 Mr Lopez's class are going on a school trip.

The cost for the trip is £9 each.

How many children handed in money today?

☐ ○ ☐ ○ ☐

☐ children handed in money today.

4 **a)** A square has a side 9 cm long.

What is the perimeter of the square?

☐ × 9 = ☐

The perimeter of the square is ☐ cm.

b) Five of these squares are put together to make another shape.

What is the perimeter of the new shape?

☐ × 9 = ☐

The perimeter of the new shape is ☐ cm.

Explain your method.

____

____

5 Which number does not divide by 9 equally? Circle the answer.

36 108 153 209 324 477

6 Rowan makes the following towers of cubes.

CHALLENGE

She now puts the cubes into towers of 9.

How many towers can she make?

____

____

Rowan can make ☐ towers of 9 cubes.

Reflect

The answer to the problem is £45 ÷ 9 = £5

What could the problem be? Write or draw your own word or picture problem to match the multiplication.

→ Textbook 4A p204

9 times-table

1 Which fact from the 9 times-table do the pictures show?

a)

☐ × ☐ = ☐

b)

☐ × ☐ = ☐

2 Use the ten frames to work out 8 × 9.

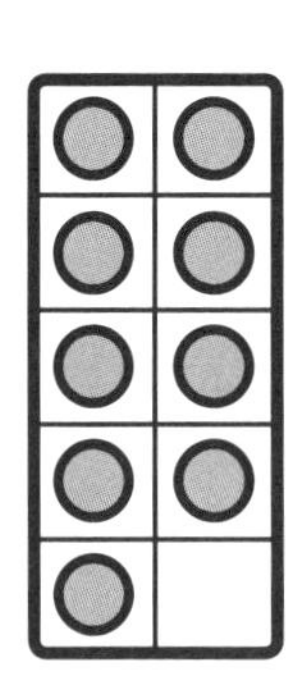
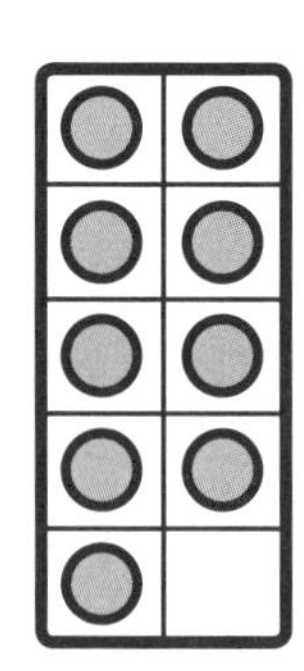

8 × 9 = ☐

3 Complete the number line.

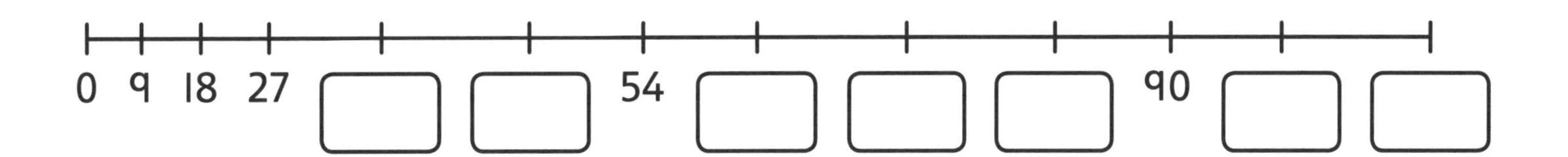

4 Complete the array to show 6 × 9.

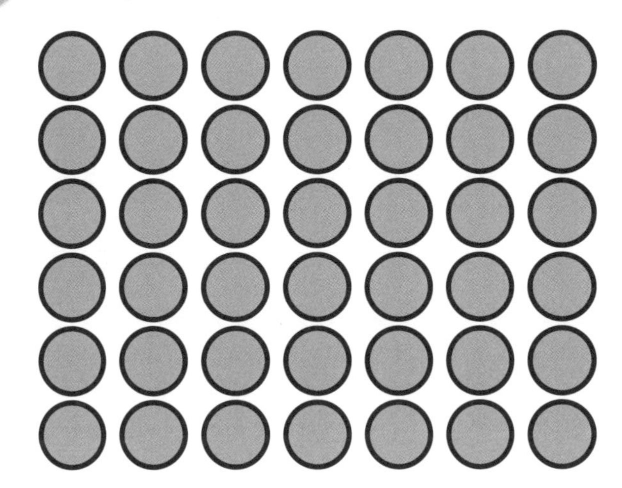

6 × 9 = ☐

5 Find the solutions to these calculations.

a) 7 × 9 = ☐

b) 0 × 9 = ☐

c) 9 × 9 = ☐

d) ☐ = 9 × 5

e) ☐ = 12 × 9

f) 18 = 2 × ☐

g) ☐ × 9 = 27

h) 9 × ☐ = 9

i) 54 ÷ 9 = ☐

j) 36 ÷ 9 = ☐

k) ☐ ÷ 9 = 11

l) ☐ ÷ 9 = 10

6 a) $3 \times 9 =$ ☐

$30 \times 9 =$ ☐

$300 \times 9 =$ ☐

$9 \times 30 =$ ☐

3 tens $\times 9 =$ ☐

b) $700 \times 9 =$ ☐

$60 \times 9 =$ ☐

$90 \times 6 =$ ☐

☐ $\times 9 = 360$

☐ $\times 9 = 9{,}000$

7 Use the digit cards to make the mathematical statements correct.

0	2	3	4	5	6	7	8

☐ $\times 9 >$ ☐ $\times 9$

☐ $\div 9 <$ ☐☐ $\div 9$

☐ $\times 9 =$ ☐☐

Reflect

In pairs, take turns to roll two dice. Multiply the total score by 9 and record your answers. How many can you get right in one minute?

→ Textbook 4A p208

Multiplying and dividing by 7

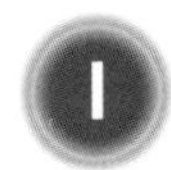

a) How many cars are parked in the car park?

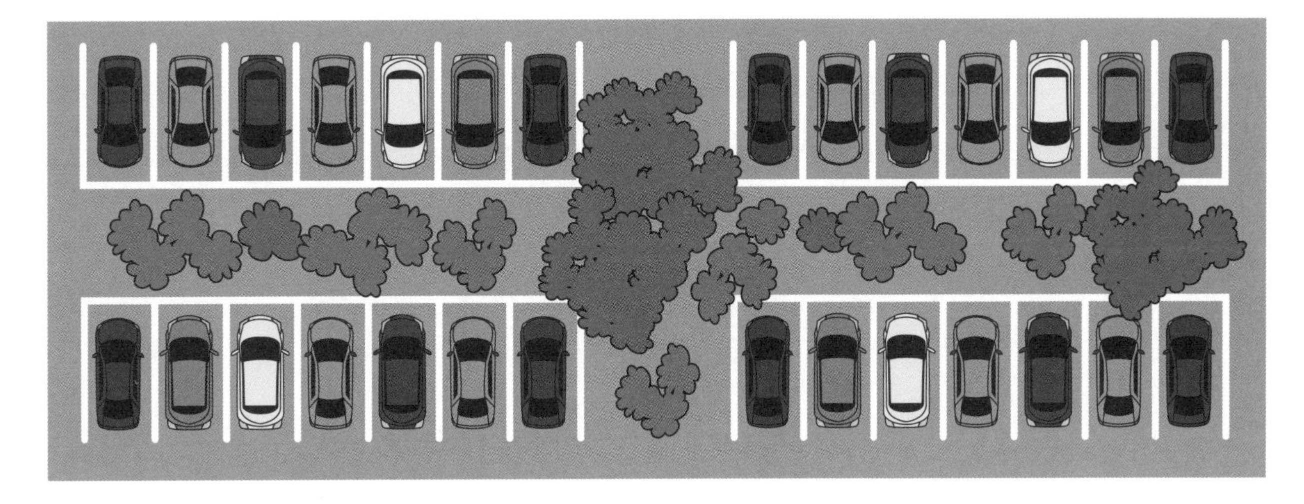

□ × □ = □

There are □ cars.

b) How many cubes are there in total?

□ × □ = □

There are □ cubes.

c) There are 7 biscuits in a packet.

Circle 49 biscuits.

2 Complete the number track.

0	7			28						

3 Draw lines to match up the correct number of days to the correct number of weeks.

4 weeks | 9 weeks | 70 weeks | 11 weeks

490 days | 63 days | 77 days | 28 days

4 **a)** 7 ten frames each have 8 counters on them.

× 7

How many counters are there in total?

☐ ◯ ☐ = ☐

There are ☐ counters in total.

b) 79 counters are put into rows of 7.

How many complete rows of counters can be formed?

How many counters are left over?

There are ☐ complete rows of counters.

There are ☐ left over.

CHALLENGE

5 Alex buys these items.

£7

£7

£7

The total cost of the items is £35.

How much does a bag of popcorn cost?

Write down your method.

First … ______________________________

____

Then … ______________________________

____

Finally … ______________________________

____

A bag of popcorn costs £[ ].

Reflect

Explain how to convince your friend that $5 \times 7 = 35$.

- ______________________________
- ______________________________
- ______________________________
-

→ Textbook 4A p212

7 times-table

1 Which facts from the 7 times-table do the pictures show?

a)

b)

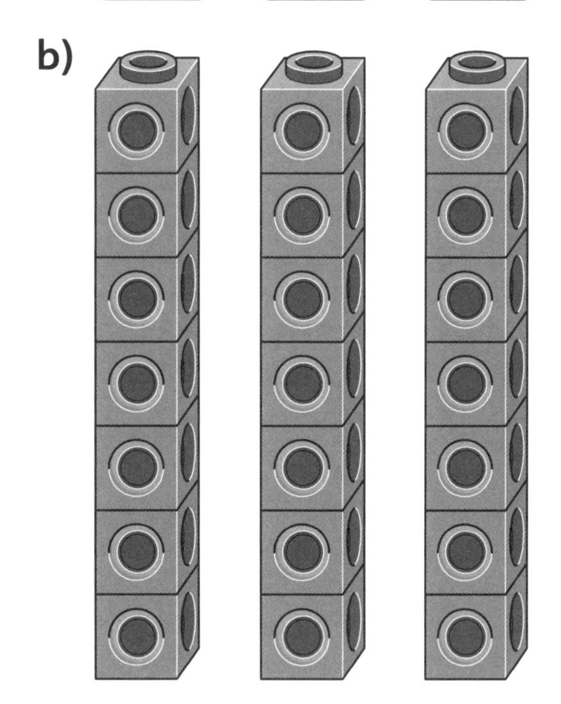

2 Complete the number line.

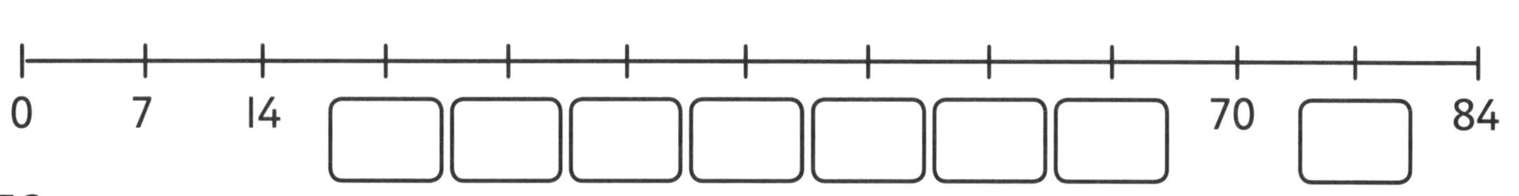

3 **a)** Alex used these cubes to work out 8 × 7.

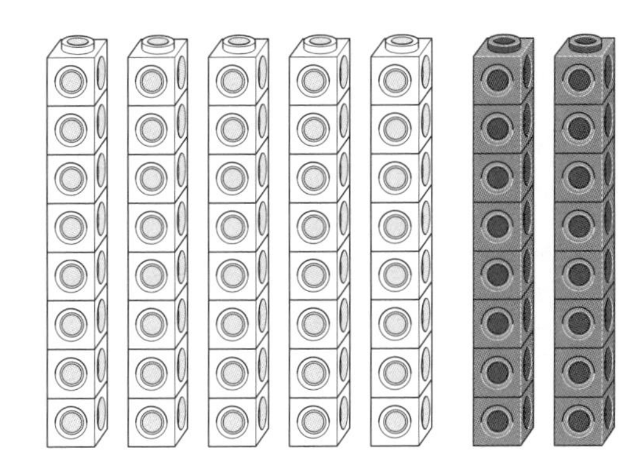

Alex's method:

First, I did 8 × 5 = ☐

Then, I did 8 × 2 = ☐

Finally, I added the numbers together: ☐ + ☐ = ☐

The answer to 8 × 7 = ☐

b) How can you use 8 × 7 to work out 9 × 7?

____

____

4 Work out the answers.

a) 4 × 7 = ☐

b) 2 × 7 = ☐

c) 7 × 5 = ☐

d) ☐ = 10 × 7

e) 7 × 0 = ☐

f) ☐ = 11 × 7

g) ☐ × 7 = 42

h) 7 × ☐ = 56

i) 77 ÷ 7 = ☐

j) 7 ÷ 7 = ☐

k) 28 ÷ 7 = ☐

l) ☐ = 63 ÷ 7

m) ☐ ÷ 7 = 3

n) ☐ ÷ 7 = 12

5 Complete the multiplication wheel.

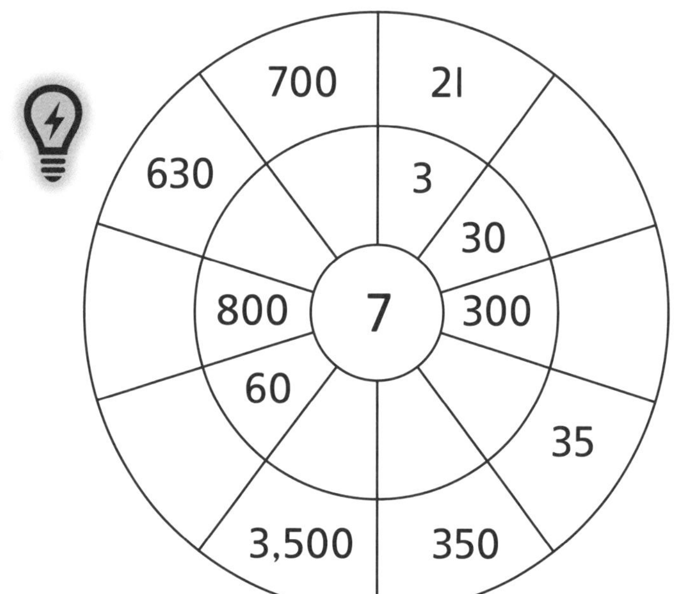

6 There are 5 areas in a zoo.

In each area there are 10 enclosures.

In each of these enclosures there are 4 animals.

Each animal has 2 eyes.

How many eyes are there in the whole zoo?

There are [ ] eyes in total.

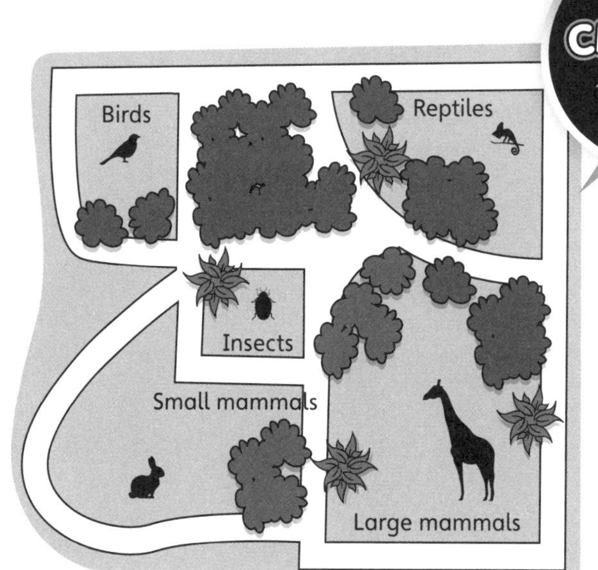

CHALLENGE

Reflect

How can you use 5 times-table and 2 times-table facts to help you work out 7 times-table facts?

→ Textbook 4A p216

11 and 12 times-tables

1 Which 11 or 12 times-table facts do the pictures show?

a)

☐ × ☐ = ☐ dots

b)

☐ × ☐ = ☐ players

2 A box contains 12 doughnuts.

How many doughnuts are in 10 boxes?

☐ × ☐ = ☐

There are ☐ doughnuts in 10 boxes.

3 Fill in the missing numbers.

a)

22	33				77			

b)

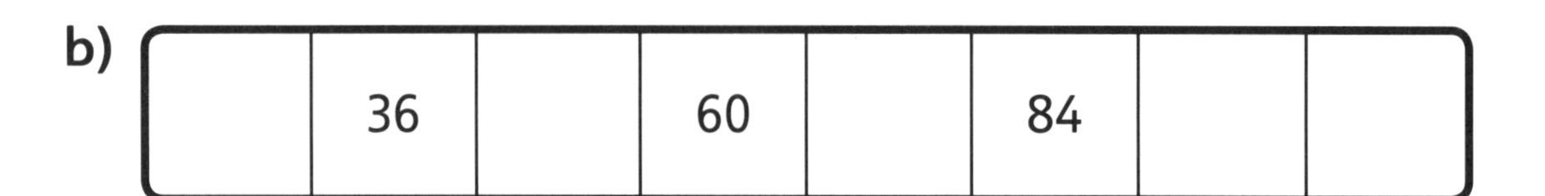

c)

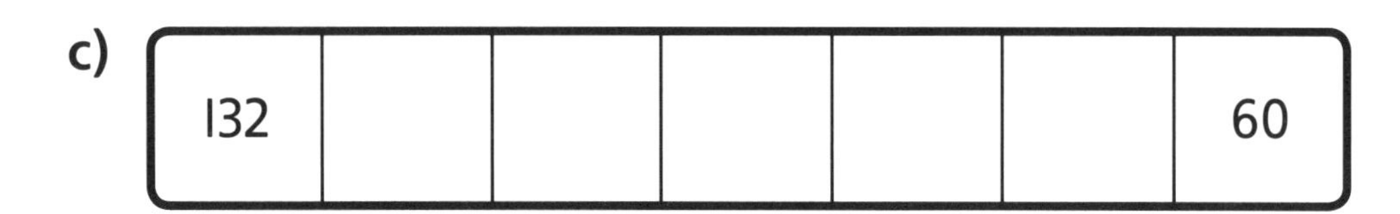

d)

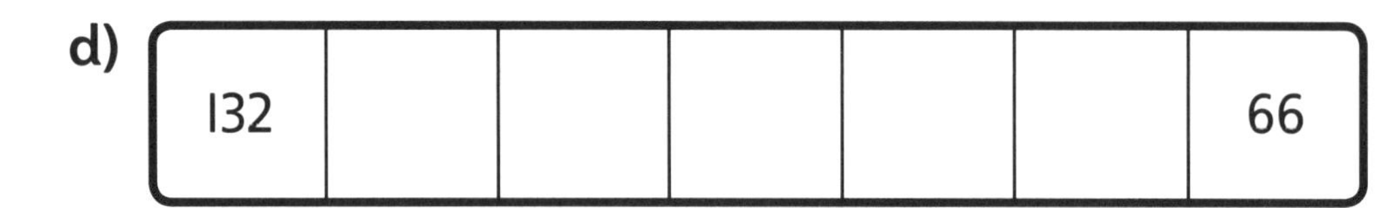

4 Complete the multiplication wheels.

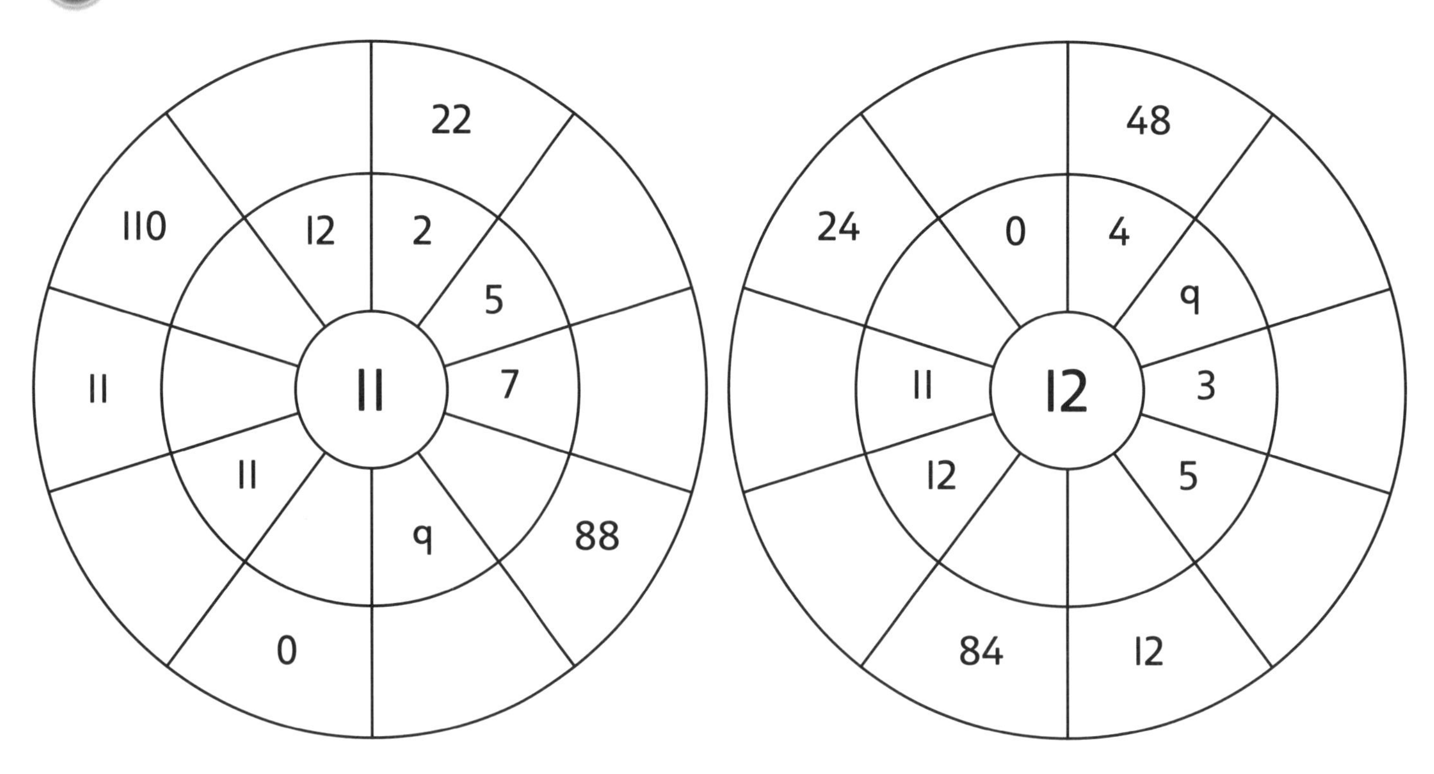

5 Complete the multiplication and division facts.

a) 6 × 12 = ☐

60 × 12 = ☐

600 × 12 = ☐

b) 880 ÷ 11 = ☐

2,400 ÷ 12 = ☐

13,200 ÷ ☐ = 1,100

13,200 ÷ ☐ = 1,200

c) ☐ = 360 ÷ 12

72 ÷ 12 = ☐ ÷ 11

9,900 ÷ 11 = ☐ × 100

Reflect

You now know all of the times-tables. Complete the grid to show off the facts you know.

×	7					11		6		1		
10												
					110				88			
						11						
4											48	
				25								
	42	18	12		60							
									16			
12										12		
7							63					
									24			
9												36
										8		

→ Textbook 4A p220

End of unit check

My journal

1 Jamilla buys some presents for her friends.

- a small present costs £3
- a medium present costs £6
- a large present costs £9

Jamilla spends £45 in total on presents.

How many of each size of present does she buy?

How many answers can you find?

2 Sort these problems into two types of calculation. Write down the reasons behind your sorting. Then work out the answers.

A: A book costs £6. How much do 7 books cost?	C: A board game costs £9. How many board games can I buy with £90?
B: A bag containing 48 sweets is shared between 6 children. How many sweets does each child receive?	D: A bag of compost weighs 18 kg. How much do 9 bags of compost weigh?

I put the problems into these groups: ______________________

The reason I sorted in this way is ______________________

____

____

Power check

How do you feel about your work in this unit?

Power puzzle

How fast can you complete each of these multiplication grids?

1 Time to complete:
____ minutes and
____ seconds

×	5	8	4	12	6	9	2	10	11	3	7	1
2												
9												
8												
12												
5												
6												
7												
3												
11												
1												
4												
10												

2 Time to complete:
____ minutes and
____ seconds

×		7		4			11		5	10		
4						4						12
						10						
										20	24	
						1						
8	16											
						6	66					
									60			
3								18				
												21
11					88							
9			81									
		35										

Design your own grid for a friend to complete. How many boxes do you need to give them before they can complete all the others?

My power points

Colour in the ☆ to show what you have learnt.

Colour in the ☺ if you feel happy about what you have learnt.

Unit I

I can ...

☆ ☺ Round numbers to the nearest I0

☆ ☺ Round numbers to the nearest I00

☆ ☺ Count in I,000s

☆ ☺ Represent 4-digit numbers

☆ ☺ Use number lines

☆ ☺ Write Roman numerals to I00

Unit 2

I can ...

☆ ☺ Find I,000 more and less

☆ ☺ Compare and order numbers to I0,000

☆ ☺ Round numbers to the nearest I,000

☆ ☺ Count in 25s

☆ ☺ Count back through 0 into negative numbers

Unit 3

I can ...

☆ ☺ Add and subtract Is, I0s, I00s and I,000s

☆ ☺ Add two 4-digit numbers using the column method

☆ ☺ Subtract two 4-digit numbers using the column method

☆ ☺ Find and use equivalent difference and other mental methods

☆ ☺ Estimate answers to additions and subtractions

☆ ☺ Check strategies

☆ ☺ Solve addition and subtraction problems

Unit 4

I can …

☆ ☺ Convert between kilometres and metres

☆ ☺ Find perimeters of shapes

☆ ☺ Work out missing lengths

☆ ☺ Solve problems involving perimeter

Unit 5

I can …

☆ ☺ Multiply by and divide multiples of 10 and 100

☆ ☺ Multiply and divide by 0 and 1

☆ ☺ Say my times-tables from 1 to 12

☆ ☺ Understand related multiplication and division facts

☆ ☺ Solve multiplication and division word problems

Keep up the good work!